Introduction to
Statistical
Mechanics
Solutions to Problems

Introduction to Statistical Mechanics

Solutions to Problems

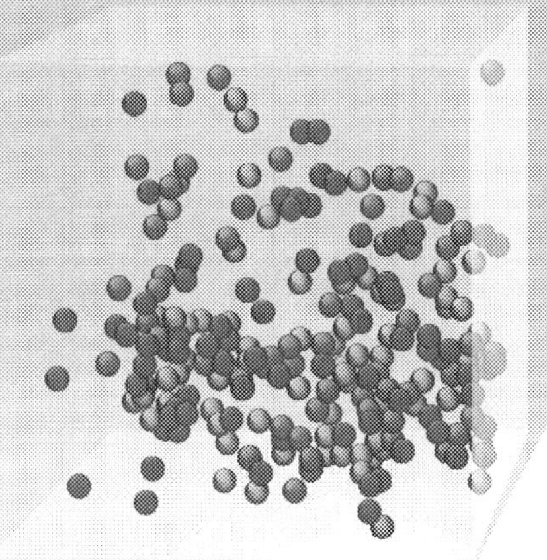

John Dirk Walecka

College of William and Mary, USA

NEW JERSEY • LONDON • SINGAPORE • BEIJING • SHANGHAI • HONG KONG • TAIPEI • CHENNAI • TOKYO

Published by

World Scientific Publishing Co. Pte. Ltd.
5 Toh Tuck Link, Singapore 596224
USA office: 27 Warren Street, Suite 401-402, Hackensack, NJ 07601
UK office: 57 Shelton Street, Covent Garden, London WC2H 9HE

Library of Congress Cataloging-in-Publication Data
Names: Walecka, John Dirk, 1932– author.
Title: Introduction to statistical mechanics : solutions to problems /
John Dirk Walecka, College of William and Mary, USA.
Description: Singapore ; Hackensack, NJ : World Scientific, [2016] |
Includes bibliographical references and index.
Identifiers: LCCN 2016035658| ISBN 9789813148130 (pbk. ; alk. paper) |
ISBN 9813148136 (pbk. ; alk. paper)
Subjects: LCSH: Statistical mechanics--Problems, exercises, etc.
Classification: LCC QC174.844 .W35 2016 | DDC 530.13--dc23
LC record available at https://lccn.loc.gov/2016035658

British Library Cataloguing-in-Publication Data
A catalogue record for this book is available from the British Library.

Printed in Singapore

Preface

The text *Introduction to Statistical Mechanics* [Walecka (2011)] is based on a one-quarter first-year graduate course I taught several times at Stanford and William and Mary. My starting point was a set of statistical assumptions due to Boltzmann. Since the counting is simpler there, the course started with quantum mechanics and then went to the classical limit. Boltzmann's approach is very powerful, for two basic statistical hypotheses allow one to successfully analyze a truly wide variety of applications. Boltzmann's approach starts with the microcanonical ensemble and microcanonical partition function, and eventually arrives at the canonical ensemble and canonical partition function, from which the Helmholtz free energy follows immediately. From there, the extension to the grand canonical ensemble and grand partition function, which yields the thermodynamic potential, takes a direct path.

I had previously published a volume entitled *Fundamentals of Statistical Mechanics: Manuscript and Notes of Felix Bloch, prepared by J. D. Walecka* [Walecka (2000)]. Bloch's work was based on Gibbs' analysis at the beginning of the last century (see [Gibbs (1960)]). Gibbs starts from classical dynamics and builds from it the statistical mechanics of many-particle assemblies. The validity of the statistical methods in Gibbs' approach depends on an appropriate dynamical behavior in phase space (it holds for "quasi-ergodic coarse-grain averaging"). Gibbs' approach again arrives at the canonical partition function for the canonical ensemble. The extension to the grand canonical ensemble and grand partition function follows as before. Quantum statistical mechanics, which came later, was introduced by Bloch through an analogous phase-space averaging. Although many examples were used to illustrate the principles, Felix always maintained that it was not a course on applications. In contrast to the Bloch course, mine was essentially all on applications. The book *Introduction to Statistical Mechanics* is meant to *complement* the Bloch book, and is in no sense a replacement for it.

The power of statistical mechanics is illustrated by the wide variety of applications covered in *Introduction to Statistical Mechanics*, including

- molecular spectroscopy
- paramagnetic and dielectric assemblies
- chemical equilibria
- normal modes in solids and the Debye model
- virial and cluster expansions for imperfect gases
- law of corresponding states
- quantum Bose and Fermi gases
- black-body radiation
- Bose-Einstein condensation
- Pauli spin paramagnetism
- Landau diamagnetism
- regular solutions
- order-disorder transitions in solids
- spin lattices and the Ising model
- U(1) lattice gauge theory

Furthermore, coverage is extended through the problems in the book to include

- white-dwarf stars
- Thomas-Fermi screening in metals
- Thomas-Fermi theory of atoms
- nuclear symmetry energy
- thermal current in metals
- keratin molecules in wool
- lattice gas
- numerical Monte Carlo and Metropolis methods
- density functional theory
- quark-gluon plasma ; *etc.*

An appendix is devoted to non-equilibrium statistical mechanics through an analysis of the Boltzmann equation and its extension to the Vlasov and Nordheim-Uehling-Uhlenbeck equations. An application to heavy-ion reactions is given there.

Many problems are included in the book, some after each chapter and the appendix. The problems are meant to enhance and extend the coverage. For the most part, the problems are not difficult, and the steps are clearly laid out. Those problems that require somewhat more effort are so noted.

Together with Paolo Amore, we have previously published solutions manuals for a three-volume series on modern physics [Amore and Walecka (2013); Amore and Walecka (2014); Amore and Walecka (2015)], the first of which was prepared at the publisher's request. Although nothing replaces a direct confrontation of the problems, these solutions manuals should provide useful teaching and learning aids for both instructors and students. In a similar fashion, the present book provides a solutions manual for the 140 problems in the text *Introduction to Statistical Mechanics* [Walecka (2011)].

That book assumes a knowledge of quantum mechanics at the level of [Walecka (2008)], and of classical mechanics at the level of [Fetter and Walecka (2003a)]. A knowledge of complex variables at the level of Appendix A of that latter reference is also assumed,[1] as well as familiarity with multi-variable calculus. The reader is assumed to have a basic knowledge of thermodynamics; however, the first chapter provides an appropriate review of that subject.

The goal of the book *Introduction to Statistical Mechanics*, and the present volume, is to provide the reader with a clear working knowledge of the very useful and powerful methods of statistical mechanics, and to enhance the understanding and appreciation of more detailed and advanced texts, such as [Fowler and Guggenheim (1949); Mayer and Mayer (1977); Tolman (1979); Landau and Lifshitz (1980); Ma (1985); Huang (1987); Chandler (1987); Kubo (1988); Negele and Ormond (1988); Kadanoff (2000); Davidson (2003); Fetter and Walecka (2003); McCoy (2015)].

Williamsburg, Virginia
February 1, 2016

John Dirk Walecka
Governor's Distinguished CEBAF
Professor of Physics, Emeritus
College of William and Mary

---

[1]The complex analysis plays a central role in the method of steepest descent used to analyze the microcanonical ensemble.

Contents

Chapter 1

Introduction

Problem 1.1 Prove from Eq. (1.1) that the integral in Eq. (1.3) is independent of path.

Solution to Problem 1.1

Equation (1.1) states the following integral around a closed cycle vanishes

$$\oint \underset{\text{Heat in}}{(đQ} - \underset{\text{Work out}}{đW)} = 0 \qquad ; \text{ All closed cycles}$$

Pick two points on the cycle (A, B) with specified thermodynamic variables. Let C_1 denote that part of the cycle running from $A \to B$, and C_2 that part of the cycle returning from $B \to A$. Then the above states

$$\int_{C_1 \hookrightarrow} (đQ - đW) + \int_{C_2 \hookrightarrow} (đQ - đW) = 0$$

where the arrows indicate the direction during the cycle. Since both the heat flow and work are algebraic and change sign if the change occurs in the opposite direction, the integral *changes sign* if the trajectory is traversed in the opposite direction[1]

$$\int_{C_2 \hookrightarrow} (đQ - đW) = -\int_{C_2 \hookleftarrow} (đQ - đW)$$

[1]To help visualize this, imagine that $\int_{C_2 \hookrightarrow}$ is along some arbitrary, fixed reversible path, and $\int_{C_1 \hookrightarrow}$ is then along *any* path connecting (A, B).

Hence the previous relation can be re-written as

$$\int_{C_1 \hookrightarrow} (đQ - đW) = \int_{C_2 \hookleftarrow} (đQ - đW)$$

This states that the integral is the same whether we go from A to B along C_1 or along C_2 in the opposite direction. Since the cycle containing the points (A, B) is arbitrary, the integral

$$\int_A^B (đQ - đW)$$

is independent of the path from $A \to B$.

Problem 1.2 Start from either statement of the second law, and see how far you can get in verifying the statements leading to Eqs. (1.5) and (1.6); then compare with [Zemansky (1968)].

Solution to Problem 1.2

The two equivalent statements of the second law of thermodynamics are given at the start of section 1.1.2:

(1) *Kelvin*: It is impossible to construct an engine that, operating in a cycle, will produce no effect other that extraction of heat from a reservoir and performance of an equivalent amount of work.
(2) *Clausius*: It is impossible to construct a device that, operating in a cycle, will produce no effect other than the transfer of heat from a cooler to a hotter body.

We shall not provide a rigorous derivation of the consequences, but leave that to a basic course in thermodynamics.[2] The key element is to establish that all reversible engines operating between two heat baths at distinct temperatures $T_1 > T_2$ have the same *efficiency*, equal to that of a Carnot engine, from which Eqs. (1.5)–(1.7) follow. The argument goes something like this. Consider *two* such engines operating between the two heat baths, and let the second operate in the reverse direction. Suppose the first absorbs heat Q_1 at T_1, produces work W, and expels heat Q_2 at T_2. Now put the heat Q_2 and work W *into* the second engine. If it does not expel exactly the same heat Q_1 into the bath at T_1, then the *combined* engines constitute a device that can be arranged to violate the second statement of the second law.

[2]Which offers a wonderful experience in scientific reasoning!

Problem 1.3 (a) Why does each point on the dotted curve in Fig. 1.2 in the text correspond to a given T?[3]

(b) How could one carry out the Carnot cycles shown in Fig. 1.2 in the text in a continuous manner with each segment being covered only once?

(c) Why is it unnecessary for the construction of the entropy to actually traverse opposing segments of the adiabats in (b)?

(d) Show that the total heat input and total work output in (b) satisfy $Q = W$.

Solution to Problem 1.3

(a) A perfect gas obeys the equation of state $PV = nRT$ (see, for example, Prob. 1.4). Thus if we specify (P, V), we also specify T;

(b) If we simply take the assembly around the outer segments of all the Carnot cycles, omitting all the common segments traversed in opposite directions, we move around the loop in a clockwise direction, in a continuous manner, with each segment being covered only once;

(c) The opposing segments along the adiabats in Fig. 1.2 in the text are traversed in a reversible manner, and along each of them the reversible heat flow vanishes, $dQ_R = 0$. Hence there is no change in entropy along those segments;

(d) Since the gas returns to its original state after completing the cycle, there is no change in internal energy. The first law then says

$$\Delta E = Q - W = 0 \qquad ; \text{ first law}$$

Problem 1.4 A perfect gas obeys the equation of state $PV = nRT = 2E(T)/3$, where n is the number of moles and R is the gas constant. Such a gas, in contact with a heat bath at temperature T, is initially confined by a thin membrane to one-half of a box of volume V. A hole is punched in the membrane so that the gas now fills the entire box.

(a) Find a reversible path, and show that the entropy change of the gas is $\Delta S = nR \ln 2$;

(b) What is the heat flow from the bath when the hole is punched?[4] What is the entropy change of the heat bath? Of the combined system of sample and heat bath?

---

[3] *Hint*: What is the equation of state of a perfect gas?

[4] *Hint*: Use the first law. Note that in this problem the reversible and irreversible processes connect two points on the *same isotherm* for the gas, $PV = nRT$.

Solution to Problem 1.4

(a) The entropy is a state function, so it is only necessary to find a reversible path between the two states to compute the change in entropy. Consider a *reversible isothermal expansion* of the gas. A combination of the first and second laws gives Eq. (1.17)

$$dE = TdS - PdV \qquad ; \text{ first and second laws}$$

Since the internal energy only depends on the temperature, it does not change. Hence

$$TdS = PdV = nRT\frac{dV}{V}$$

Here the equation of state $P = nRT/V$ has been employed in obtaining the second equality. Integration then gives

$$\Delta S = nR\ln\frac{V_2}{V_1} = nR\ln 2$$

(b) When the hole is punched and the gas fills the box, one has an irreversible process. The first law states that

$$\Delta E = \Delta Q - \Delta W \qquad ; \text{ first law}$$

Since the gas remains at the temperature T, one has $\Delta E = 0$, and since no external work is done, $\Delta W = 0$. Therefore, there is no heat flow

$$\Delta Q = 0$$

The entropy change of the heat bath thus vanishes. We conclude that in this irreversible process, the following inequality is satisfied for the combined system of sample and heat bath

$$\frac{\Delta Q}{T} - \Delta S \leq 0$$

Problem 1.5 N objects of spin 1/2 sit on distinct, localized sites, and the assembly is unpolarized.

(a) Show the spin entropy of this system is $S = Nk_{\rm B}\ln 2$;

(b) At a temperature T the spins are observed to align. Show that an amount of heat $Q = Nk_{\rm B}T\ln 2$ must have been extracted to produce this configuration;

(c) What is the corresponding change in internal spin energy?

Solution to Problem 1.5

(a) If N objects of spin 1/2 sit on distinct, localized sites, and the assembly is unpolarized, then the total number of complexions is

$$\Omega = 2^N$$

Hence the spin entropy is

$$S = k_{\rm B} \ln 2^N = N k_{\rm B} \ln 2$$

(b) If the spins are all aligned, then the number of complexions is $\Omega = 1$, and the spin entropy vanishes. Hence the increase in entropy as the spins depolarize at a temperature T is

$$\Delta S = N k_{\rm B} \ln 2$$

The corresponding reversible heat flow to the assembly as the spins depolarize then follows from the second law

$$Q = T\Delta S = N k_{\rm B} T \ln 2$$

(c) If no external work is done during the depolarization, then the first law implies that the corresponding increase in internal spin energy is

$$\Delta E = Q = N k_{\rm B} T \ln 2$$

Problem 1.6 (a) A piston under a pressure P expands quasistatically. Show that the reversible work done in the surroundings is $dW = PdV$;

(b) Show that the second law implies that the reversible (quasistatic) heat flow to a sample at an absolute temperature T is $dQ = TdS$.

Solution to Problem 1.6

(a) A piston under a pressure P expands quasistatically as in Fig. 1.1. For a quasistatic process, the pressure exerted by the assembly on the piston just balances the external pressure P.[5] Hence the force exerted by the piston is

$$F = PA$$

The external work done *by* the piston is the force times the distance through which it moves. For displacement through an infinitesimal distance dx, this

[5] It is actually infinitesimally higher.

becomes

$$dW = PA\,dx = PdV \qquad ;\ \text{external work}$$

where $dV = Adx$ is the increase in volume of the assembly.

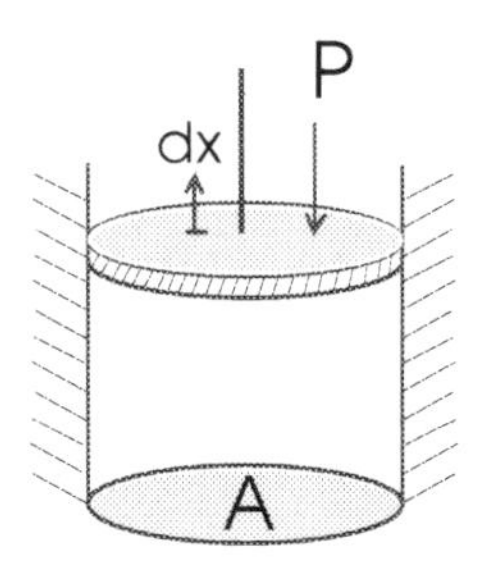

Fig. 1.1 A piston under a pressure P expands quasistatically.

(b) If the reversible heat flow *to* the assembly during the quasistatic expansion is dQ_R, then the second law gives

$$dS = \frac{dQ_R}{T} \qquad ;\ \text{second law}$$

A combination of the first and second laws than gives

$$dE = TdS - PdV \qquad ;\ \text{first and second laws}$$

Problem 1.7 Use an argument similar to that given in the text for the Helmholtz free energy to derive Gibbs criterion in Eq. (1.34) for equilibrium in a sample at fixed (P, T).

Solution to Problem 1.7

The Gibbs free energy is a state function defined by

$$G = E + PV - TS$$

It follows, as in the first of Eqs. (1.33), that for a transition at fixed temperature and pressure

$$(\delta G)_{T,P} = \delta E + P\delta V - T\delta S$$

As in the first of Eqs. (1.31), the first law states

$$\delta E = đQ - P\delta V$$

Therefore, for a transition at fixed temperature and pressure

$$(\delta G)_{T,P} = đQ - T\delta S$$

Now, just as in the second of Eqs. (1.33), for an allowable transition this quantity will decrease

$$(\delta G)_{T,P} \leq 0$$

Hence, Gibbs criterion for *equilibrium* is

$$(\delta G)_{T,P} \geq 0 \qquad ; \text{ Gibbs criterion}$$

At fixed (T, P), the equilibrium state of a sample will be the one of *minimum Gibbs free energy.*

Problem 1.8 The *enthalpy* is a state function that is useful at fixed pressure. It is defined by making the following Legendre transformation

$$\mathcal{H} \equiv E + PV \qquad ; \text{ enthalpy}$$

(a) Show that with pressure-volume work, the first law of thermodynamics becomes

$$d\mathcal{H} = đQ + VdP \quad ; \text{ first law}$$

(b) Show that for reversible (quasistatic) processes, the first and second laws become

$$d\mathcal{H} = TdS + VdP \qquad ; \text{ first and second law}$$

(c) Show that Gibbs criterion for equilibrium takes the form

$$(\delta\mathcal{H})_{S,P} \geq 0 \qquad ; \text{ Gibbs criterion}$$

A sample in equilibrium at fixed (S, P) will minimize its *enthalpy.*

Solution to Problem 1.8

(a) The enthalpy is a state function useful in chemistry and engineering where, in contrast to processes taking place inside a box at fixed volume, those processes take place in a container whose volume can change at a fixed (often atmospheric) pressure. The enthalpy is defined by

$$\mathcal{H} \equiv E + PV \qquad ; \text{ enthalpy}$$

With pressure-volume work $dW = PdV$, and therefore the first law $dE = đQ - PdV$ implies

$$\begin{aligned} d\mathcal{H} &= dE + PdV + VdP \\ &= đQ + VdP \qquad ; \text{ first law} \end{aligned}$$

(b) For a reversible (quasistatic) processes, the second law says $dQ_R = TdS$. Therefore, the first and second laws become

$$d\mathcal{H} = TdS + VdP \qquad ; \text{ first and second law}$$

(c) The derivation of Gibbs criterion for equilibrium now follows exactly as in Eqs. (1.28)–(1.31) in the text. The result is

$$(\delta\mathcal{H})_{S,P} \geq 0 \qquad ; \text{ Gibbs criterion}$$

Problem 1.9 (a) Consider a sample of volume V, surface temperature T, and surface pressure P. Divide it into tiny subunits. Introduce a heat flow dQ_R distributed over the surface, and show $\delta S = dQ_R/T$;[6]

(b) Now introduce *any* heat flow $đQ$ distributed over the surface, and derive Eq. (1.30);

(c) Derive the first of Eqs. (1.31).

Solution to Problem 1.9

(a) In the discussion of thermodynamic equilibrium in the text, the concept of a variation δ in a state function is introduced, where, for example, *δS in Eq. (1.28) now means any possible variation, not necessarily one leading to an equilibrium state.* An example of such a variation is shown in Fig. 1.4 in the text.[7] To determine δS for the sample, one divides it into infinitesimal subunits, computes dS_i for that subunit, and then sums the

[6] It is assumed here that any subsequent reversible heat flow across the surface of *interior* subunits cancels in this argument (compare the discussion of entropy conservation in chapter 60 of [Fetter and Walecka (2003a)]).

[7] The differential dS denotes an infinitesimal difference *between equilibrium states.*

results for the sample

$$\delta S = \sum_i dS_i$$

Consider a sample of volume V, surface temperature T, and surface pressure P. Divide it into tiny subunits containing a given amount of material.[8] Imagine that there is a reversible heat flow dQ_R distributed over the surface of a sample. Then for each subunit on the surface

$$dS_i = \frac{dQ_{R\,i}}{T} \qquad ; \text{ subunit}$$

With the assumption that any reversible heat flow in and out of the interior subunits cancels when summed over the sample, the change in entropy of the sample will be

$$\delta S = \frac{dQ_R}{T} \qquad ; \text{ sample}$$

(b) Now introduce *any* heat flow $đQ$ distributed over the surface of the sample. For each subunit on the surface

$$dS_i - \frac{đQ_i}{T} \geq 0 \qquad ; \text{ subunit}$$

Any internal heat flow in and out of the interior subunits again cancels, and hence when summed over the subunits the variation in entropy of the sample is[9]

$$\delta S - \frac{đQ}{T} \geq 0 \qquad ; \text{ sample}$$

This is Eq. (1.30), which expresses the second law of thermodynamics.

(c) Similar arguments can be applied to the internal energy E. The internal heat flow between subunits cancels when summed over subunits, and the net heat flow to the sample is $đQ$. The pressure-volume work cancels locally across any small common surface area of the subunits; however, when summed over the subunits there is a net volume change of δV, and with a surface pressure P, there is net external work of $P\delta V$. Therefore

$$\delta E = đQ - P\delta V \qquad ; \text{ sample}$$

[8] Note that the internal (T, P) can vary in these arguments.

[9] Any internal viscous dissipation only increases δS in this relation (see chapter 60 of [Fetter and Walecka (2003a)]).

This is the first of Eqs. (1.31), which expresses energy conservation and the first law.

These relations provide the basis for the discussion of thermodynamic equilibrium in the text.

Problem 1.10 Given an assembly of N localized systems with two energy levels separated by ε. Suppose the systems are initially all in the excited state and the temperature satisfies $k_{\rm B}T \ll \varepsilon$.

(a) What is the initial entropy of the assembly? What is the entropy change ΔS to the state of the assembly where the systems are all in the ground state? What is the energy change ΔE?

(b) Use the finite form of the stability criterion in Eq. (1.29), $(\Delta E)_{S,V} \geq 0$, to show the initial state of this assembly is *unstable*;

(c) What is the heat flow in the transition to the final state? Show that the inequality $\Delta Q/T - \Delta S \leq 0$ is satisfied.

Solution to Problem 1.10

(a) If the systems are initially all in the excited state, then there is just one complexion $\Omega = 1$, and the initial entropy vanishes. If the systems are all in the ground state when the assembly is placed in contact with a heat bath at a temperature $k_{\rm B}T \ll \varepsilon$, then the final entropy also vanishes, and hence the change in entropy of the assembly is

$$\Delta S = S_f - S_i = 0$$

The initial internal energy is $E_i = N\varepsilon$, and the final energy is $E_f = 0$, thus the change in internal energy of the assembly is

$$\Delta E = E_f - E_i = -N\varepsilon$$

(b) The finite form of the stability criterion in Eq. (1.29) is

$$(\Delta E)_{S,V} \geq 0 \qquad ; \text{ Gibbs criterion}$$

Here

$$(\Delta E)_{S,V} = -N\varepsilon \qquad ; \text{ unstable}$$

Hence the initial configuration is *unstable*, and the systems will indeed all transition to the ground state.

(c) If no external work is done during the transition, then the first law says the heat flow *to* the assembly is

$$\Delta Q = \Delta E = -N\varepsilon$$

It follows that

$$\frac{\Delta Q}{T} - \Delta S = -\frac{N\varepsilon}{T}$$

Hence in this process the assembly indeed satisfies the following inequality

$$\frac{\Delta Q}{T} - \Delta S \leq 0$$

Chapter 2

The Microcanonical Ensemble

Problem 2.1 Construct the complexions for 3 excited systems on 5 sites and show the total number of complexions is $\Omega = 5!/3!\,2! = 10$.

Solution to Problem 2.1

In Fig. 2.1 we plot the complexions with three excited systems, denoted with crosses, on five sites, labeled with dots. As stated, the number of complexions is $\Omega = 5!/3!\,2! = 10$.

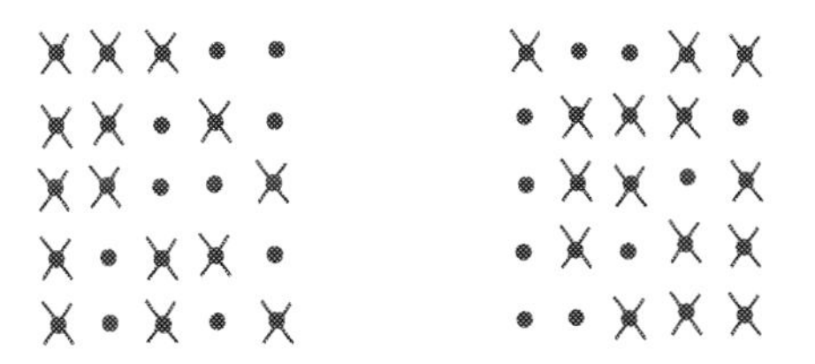

Fig. 2.1 Complexions with three excited systems, denoted with crosses, on five sites, labeled with dots.

Problem 2.2 Separate an analytic function into its real and imaginary parts $f(z) = u(x, y) + iv(x, y)$. An analytic function has a derivative that is independent of the direction in which it is taken in the complex plane. Evaluate the derivative first in the x-direction and then in the y-direction to show

$$\frac{df}{dz} = \frac{\partial u}{\partial x} + i\frac{\partial v}{\partial x} = \frac{\partial u}{\partial iy} + i\frac{\partial v}{\partial iy}$$

Hence deduce the *Cauchy-Riemann equations*

$$\frac{\partial u}{\partial x} = \frac{\partial v}{\partial y} \quad ; \quad \frac{\partial u}{\partial y} = -\frac{\partial v}{\partial x}$$

Solution to Problem 2.2

The complex variable is $z = x + iy$. An analytic $f(z)$ with $f(z) = u(x, y) + iv(x, y)$, where (u, v) are real functions of (x, y), has a derivative that is independent of direction in the complex plane. Evaluate that derivative first in the horizontal direction at fixed y, so that $dz = dx$

$$\frac{df}{dz} = \frac{\partial f}{\partial x} = \frac{\partial u}{\partial x} + i\frac{\partial v}{\partial x}$$

Here the partial derivative indicates that y is to be held fixed.

Now evaluate the derivative with respect to z in the vertical direction at fixed x, with $dz = idy$

$$\frac{df}{dz} = \frac{1}{i}\frac{\partial f}{\partial y} = \frac{1}{i}\frac{\partial u}{\partial y} + \frac{\partial v}{\partial y}$$

Here the partial derivative indicates that x is to be held fixed.

The expressions for df/dz must be equal, and equating the real and imaginary parts then gives

$$\frac{\partial u}{\partial x} = \frac{\partial v}{\partial y} \quad ; \quad \frac{\partial u}{\partial y} = -\frac{\partial v}{\partial x}$$

These are the *Cauchy-Riemann equations* satisfied by the real and imaginary parts of an analytic function.

Problem 2.3 The modulus of the integrand in Eq. (2.71) is defined in terms of its real and imaginary parts by $|I(z)| = \sqrt{u^2 + v^2}$. Along the real axis, $I(z)$ is real so that $|I| = u$ (positive) and $v = 0$.

(a) The condition that the integrand has a minimum in the x-direction at $x = x_0$ is then $[\partial u/\partial x]_{x_0} = [\partial v/\partial x]_{x_0} = 0$. Show this implies $[\partial |I|/\partial x]_{x_0} = 0$;

(b) Use the Cauchy-Riemann equations of Prob. 2.2 to show that $[\partial v/\partial y]_{x_0} = [\partial u/\partial y]_{x_0} = 0$. Hence show that $[\partial |I|/\partial y]_{x_0} = 0$;

(c) Use these results to show

$$\left[\frac{\partial^2 |I|}{\partial x^2}\right]_{x_0} = \left[\frac{\partial^2 u}{\partial x^2}\right]_{x_0} \quad ; \quad \left[\frac{\partial^2 |I|}{\partial y^2}\right]_{x_0} = \left[\frac{\partial^2 u}{\partial y^2}\right]_{x_0}$$

Hence conclude that

$$\left[\left(\frac{\partial^2}{\partial x^2} + \frac{\partial^2}{\partial y^2}\right)|I|\right]_{x_0} = \left[\left(\frac{\partial^2}{\partial x^2} + \frac{\partial^2}{\partial y^2}\right)u\right]_{x_0}$$

(d) Show $|\,[f(z)]^N/z^E\,| = |f(z)|^N/|z|^E$.

Solution to Problem 2.3

(a) The function $I(z) = u(x,y) + iv(x,y)$ defined in Eq. (2.71) and illustrated in Fig. 2.5 in the text is analytic in z, real along the real axis with $v(x,0) = 0$, and has a minimum along the real axis at a point $(x,y) = (x_0, 0)$. Thus if we take derivatives along the real axis

$$\left[\frac{\partial u}{\partial x}\right]_{x_0} = \left[\frac{\partial v}{\partial x}\right]_{x_0} = 0$$

The modulus of $I(z)$ is defined as

$$|I(z)| \equiv \sqrt{u^2 + v^2}$$

This quantity has a saddle point at $(x,y) = (x_0, 0)$ as illustrated in Fig. 2.6 in the text, and we study the behavior of the modulus in the vicinity of the saddle point. Take the derivative along the real axis

$$\left[\frac{\partial}{\partial x}|I(z)|\right]_{x_0} = \left[\frac{\partial u}{\partial x}\right]_{x_0} = 0$$

(b) Now take the derivative in the y-direction, and use the Cauchy-Riemann equations in Prob. 2.2

$$\begin{aligned}\frac{\partial}{\partial y}|I(z)| &= \frac{1}{\sqrt{u^2+v^2}}\left[u\frac{\partial u}{\partial y} + v\frac{\partial v}{\partial y}\right] \\ &= \frac{1}{\sqrt{u^2+v^2}}\left[-u\frac{\partial v}{\partial x} + v\frac{\partial u}{\partial x}\right]\end{aligned}$$

Evaluation at $(x_0, 0)$ gives

$$\left[\frac{\partial}{\partial y}|I(z)|\right]_{x_0} = 0$$

Thus the slope of $|I(z)|$ vanishes in both directions at $(x_0, 0)$.

(c) Take a second derivative along the x-axis

$$\left[\frac{\partial^2}{\partial x^2}|I(z)|\right]_{x_0} = \left[\frac{\partial^2 u}{\partial x^2}\right]_{x_0}$$

Now take a second derivative of the relation in part (b) in the y-direction

$$\frac{\partial^2}{\partial y^2}|I(z)| = \frac{-1}{(u^2+v^2)^{3/2}}\left[u\frac{\partial u}{\partial y} + v\frac{\partial v}{\partial y}\right]^2 + \frac{1}{\sqrt{u^2+v^2}}\left[u\frac{\partial^2 u}{\partial y^2} + \left(\frac{\partial u}{\partial y}\right)^2 + v\frac{\partial^2 v}{\partial y^2} + \left(\frac{\partial v}{\partial y}\right)^2\right]$$

Use the Cauchy-Riemann equations once more

$$\frac{\partial^2}{\partial y^2}|I(z)| = \frac{-1}{(u^2+v^2)^{3/2}}\left[-u\frac{\partial v}{\partial x} + v\frac{\partial u}{\partial x}\right]^2 + \frac{1}{\sqrt{u^2+v^2}}\left[u\frac{\partial^2 u}{\partial y^2} + \left(\frac{\partial v}{\partial x}\right)^2 + v\frac{\partial^2 v}{\partial y^2} + \left(\frac{\partial u}{\partial x}\right)^2\right]$$

Evaluate this at $(x_0, 0)$. The only remaining non-zero term is

$$\left[\frac{\partial^2}{\partial y^2}|I(z)|\right]_{x_0} = \left[\frac{\partial^2 u}{\partial y^2}\right]_{x_0}$$

Hence we conclude that for the function $I(z)$ at the saddle point $(x_0, 0)$, the laplacian applied to its modulus is equal to the laplacian applied to its real part

$$\left[\left(\frac{\partial^2}{\partial x^2} + \frac{\partial^2}{\partial y^2}\right)|I|\right]_{x_0} = \left[\left(\frac{\partial^2}{\partial x^2} + \frac{\partial^2}{\partial y^2}\right)u\right]_{x_0}$$

(d) Any complex number can be represented by $z = \rho e^{i\phi}$, where ρ is the modulus and ϕ the phase. Hence

$$z = \rho e^{i\phi} \qquad ; \; f(z) = \rho_f \, e^{i\phi_f}$$

It follows that

$$\frac{[f(z)]^N}{z^E} = \frac{\rho_f^N \, e^{iN\phi_f}}{\rho^E \, e^{iE\phi}}$$

Now take the modulus of this expression. This gives

$$\left|\frac{[f(z)]^N}{z^E}\right| = \frac{\rho_f^N}{\rho^E} = \frac{|f(z)|^N}{|z|^E}$$

which is the desired result.

Problem 2.4 (a) Write $z = x_0 + iy$, expand in y, and verify the statement that the $1/z$ in the integrand in Eq. (2.91) can be evaluated at the saddle point as $1/x_0$ to $O(1/N)$;

(b) Verify that the error incurred in extending the limits on the y-integral to $\pm\infty$ in Eq. (2.82) is covered by the stated error in Eq. (2.91);

(c) Explain the appropriate limiting process used in arriving at the error estimate in Eq. (2.93).

Solution to Problem 2.4

(a) Start from the modified form of Eq. (2.82) with the $1/z$ in the integrand

$$\begin{aligned}\Omega(E,N) &= \frac{1}{2\pi i}\oint \frac{dz}{z}\, e^{Ng(z)} \\ &\approx \frac{1}{2\pi} e^{Ng(x_0)} \int_{-\infty}^{\infty} \frac{dy}{x_0 + iy}\, e^{-\frac{1}{2}Ng''(x_0)y^2} \qquad ; \; z = x_0 + iy\end{aligned}$$

Now expand in y/x_0

$$\Omega(E,N) = \frac{1}{2\pi x_0} e^{Ng(x_0)} \int_{-\infty}^{\infty} dy \left(1 - \frac{iy}{x_0} - \frac{y^2}{x_0^2} + \cdots\right) e^{-\frac{1}{2}Ng''(x_0)y^2}$$

The linear correction is odd in y and integrates to zero. For the quadratic term

$$\int_{-\infty}^{\infty} y^2\, dy\, e^{-\frac{1}{2}Ng''(x_0)y^2} = \frac{1}{[Ng''(x_0)/2]^{3/2}} \int_{-\infty}^{\infty} t^2 dt\, e^{-t^2}$$

This is down by $O(1/N)$ from the leading term in Eq. (2.91). Thus

$$\Omega(E,N) = \frac{1}{x_0}\left\{\frac{e^{Ng(x_0)}}{2\pi[Ng''(x_0)/2]^{1/2}} \int_{-\infty}^{\infty} dt\, e^{-t^2}\right\}\left[1 + O\left(\frac{1}{N}\right)\right]$$

Hence the $1/z$ can indeed be evaluated at x_0 to $O(1/N)$.

(b) Suppose we integrate over the peak in Fig. 2.7 in the text from $-y_0$ to y_0, so that Eq. (2.82) reads

$$\begin{aligned}\Omega(E,N) &= \frac{1}{2\pi i}\oint e^{Ng(z)} \\ &\approx \frac{1}{2\pi} e^{Ng(x_0)} \int_{-y_0}^{y_0} dy\, e^{-\frac{1}{2}Ng''(x_0)y^2}\end{aligned}$$

Now rewrite this identically as

$$\Omega(E,N) = \frac{1}{2\pi} e^{Ng(x_0)} \frac{1}{[Ng''(x_0)/2]^{1/2}}$$
$$\times \left[\int_{-\infty}^{\infty} e^{-t^2}\, dt - \int_{-\infty}^{-y_0[Ng''(x_0)/2]^{1/2}} e^{-t^2}\, dt - \int_{y_0[Ng''(x_0)/2]^{1/2}}^{\infty} e^{-t^2}\, dt \right]$$

Since the integrands are exponentially small, both corrections are certainly smaller than $O(1/N)$ relative to the leading term. Therefore

$$\frac{1}{2\pi i} \oint e^{Ng(z)} = \frac{e^{Ng(x_0)}}{2\pi[Ng''(x_0)/2]^{1/2}} \int_{-\infty}^{\infty} e^{-t^2}\, dt \left[1 + O\left(\frac{1}{N}\right)\right]$$

(c) The problem is the sign of the exponent in the correction term, since $(z-x_0)^4 = +y^4$. We must keep the limits of the y-integration finite until the end. The Taylor series expansion in the exponent then gives

$$\int_{iy_0}^{iy_0} dz\, e^{Ng(z)} = ie^{Ng(x_0)} \int_{-y_0}^{y_0} dy$$
$$\times \exp\left\{ -\frac{N}{2!} g''(x_0)y^2 - i\frac{N}{3!} g'''(x_0)y^3 + \frac{N}{4!} g''''(x_0)y^4 + \cdots \right\}$$

The correction from the y^3 term is odd and integrates to zero. Thus[1]

$$\int_{iy_0}^{iy_0} dz\, e^{Ng(z)} \approx \frac{ie^{Ng(x_0)}}{[Ng''(x_0)/2]^{1/2}} \int_{-t_0}^{t_0} e^{-t^2} dt\, \exp\left\{ \frac{g''''(x_0)}{6N[g''(x_0)]^2} t^4 \right\}$$

Now expand

$$\exp\left\{ \frac{g''''(x_0)}{6N[g''(x_0)]^2} t^4 \right\} = 1 + \frac{g''''(x_0)}{6N[g''(x_0)]^2} t^4 + \cdots$$

It follows that

- For finite t_0, the correction to the integral is explicitly of $O(1/N)$ relative to the leading term;
- As in part (b), the limit $t_0 \to \infty$ can then be safely taken in the correction term.

Hence, again,

$$\frac{1}{2\pi} \int_{iy_0}^{iy_0} dz\, e^{Ng(z)} = \frac{ie^{Ng(x_0)}}{2\pi[Ng''(x_0)/2]^{1/2}} \int_{-\infty}^{\infty} e^{-t^2}\, dt \left[1 + O\left(\frac{1}{N}\right)\right]$$

[1] Here $t_0 = [Ng''(x_0)/2]^{1/2} y_0$. Note the neglected term can also contribute to $O(1/N)$.

Problem 2.5 Show from Eq. (2.100) that the method of steepest descent reproduces the familiar Boltzmann distribution $n_i^\star/N = e^{\beta\varepsilon_i}/\sum_i e^{\beta\varepsilon_i}$ for the occupation numbers.

Solution to Problem 2.5

The Boltzmann distribution is derived in Eqs. (2.41)–(2.42)

$$N = \sum_i n_i^\star \qquad ; \; E = \sum_i \varepsilon_i n_i^\star$$
$$\frac{n_i^\star}{N} = \frac{e^{-\varepsilon_i/k_\mathrm{B}T}}{\sum_i e^{-\varepsilon_i/k_\mathrm{B}T}}$$

The relation locating the saddle point in the method of steepest descent in Eq. (2.100) is

$$\frac{E}{N} = \frac{\sum_i \varepsilon_i e^{\beta\varepsilon_i}}{\sum_i e^{\beta\varepsilon_i}} \qquad ; \; \beta = -\frac{1}{k_\mathrm{B}T}$$

where it is subsequently shown in Eqs. (2.103) that $\beta = -1/k_\mathrm{B}T$. Define the distribution numbers $n_i^\star$ in this case through the relations

$$N = \sum_i n_i^\star \qquad ; \; E = \sum_i \varepsilon_i n_i^\star$$
$$\frac{E}{N} = \frac{\sum_i \varepsilon_i n_i^\star}{\sum_i n_i^\star}$$

Then, just as in the Boltzmann case, the distribution numbers satisfy

$$\frac{n_i^\star}{N} = \frac{e^{-\varepsilon_i/k_\mathrm{B}T}}{\sum_i e^{-\varepsilon_i/k_\mathrm{B}T}}$$

Problem 2.6 Use the microcanonical ensemble in equilibrium statistical mechanics to derive the following results:

(a) The Maxwell-Boltzmann distribution of velocities in an ideal gas

$$\frac{\Delta n}{n} = \frac{4\pi v^2 \Delta v}{(2\pi k_\mathrm{B}T/m)^{3/2}} \exp\left(-\frac{mv^2}{2k_\mathrm{B}T}\right)$$

(b) Halley's formula for the density distribution in an isothermal atmosphere

$$\rho = \rho_0 \exp\left(-\frac{mgh}{k_\mathrm{B}T}\right)$$

(c) The concentration distribution of macro-molecules in solution in an ultracentrifuge

$$c = c_0 \exp\left(\frac{mr^2\omega^2}{2k_{\rm B}T}\right)$$

Solution to Problem 2.6

The various answers in this problem all follow from the Boltzmann distribution in Eqs. (2.42)–(2.43)

$$\frac{n_i^\star}{N} = \frac{e^{-\varepsilon_i/k_{\rm B}T}}{\sum_i e^{-\varepsilon_i/k_{\rm B}T}} \qquad ; \quad \frac{n_i^\star}{n_j^\star} = \frac{e^{-\varepsilon_i/k_{\rm B}T}}{e^{-\varepsilon_j/k_{\rm B}T}}$$

(a) In an ideal gas, the kinetic energy is

$$\varepsilon = \frac{1}{2}mv^2$$

There will be a continuous distribution of velocities

$$\frac{dn}{N} = \frac{e^{-mv^2/2k_{\rm B}T}\,4\pi v^2 dv}{\int_0^\infty e^{-mv^2/2k_{\rm B}T}\,4\pi v^2 dv}$$

where $4\pi v^2 dv$ is the volume element. The integral of the distribution is[2]

$$\begin{aligned}\int_0^\infty e^{-mv^2/2k_{\rm B}T}\,4\pi v^2 dv &= 4\pi\left(\frac{2k_{\rm B}T}{m}\right)^{3/2}\int_0^\infty x^2 e^{-x^2}\,dx \\ &= \left(\frac{2\pi k_{\rm B}T}{m}\right)^{3/2}\end{aligned}$$

Hence the normalized distribution, with finite intervals $(\Delta n, \Delta v)$, is

$$\frac{\Delta n}{n} = \frac{4\pi v^2 \Delta v}{(2\pi k_{\rm B}T/m)^{3/2}} \exp\left(-\frac{mv^2}{2k_{\rm B}T}\right)$$

This is the Maxwell-Boltzmann distribution of velocities in an ideal gas.

(b) In the atmosphere, there is a gravitational potential at a height h above the earth's surface of

$$\phi_g = gh \qquad ; \ \mathbf{f}_g = -\boldsymbol{\nabla}\phi_g = -g\,\mathbf{e}_z$$

[2]Use $\int_0^\infty x^2 e^{-x^2}\,dx = \sqrt{\pi}/4$ (see Prob. 2.14).

where $\mathbf{f}_g$ is the force per unit mass, and $\mathbf{e}_z$ is a unit vector in the upward z-direction. Hence, there is a gravitational potential energy of

$$\varepsilon_g = mgh$$

The number of systems, reflected in the particle mass density, is then distributed in height according to

$$\frac{\rho}{\rho_0} = e^{-mgh/k_{\rm B}T}$$

where ρ_0 is the density at the surface.[3] This is Halley's formula for the density distribution in an isothermal atmosphere.

(c) Go to the rest frame of the fluid in the centrifuge. In this frame, rotating with an angular velocity ω, there will be an effective centrifugal force with an effective potential

$$\phi_c = -\frac{1}{2}\omega^2 r^2 \qquad ; \; \mathbf{f}_c = -\boldsymbol{\nabla}\phi_c = \omega^2 r\, \mathbf{e}_r$$

where $\mathbf{f}_c$ is the force per unit mass, and $\mathbf{e}_r$ is a unit vector in the outward radial direction. Hence, there will be an effective centrifugal potential energy

$$\varepsilon_c = -\frac{1}{2}m\omega^2 r^2$$

It follows that there will be a distribution in number of systems, reflected in the concentration, of

$$\frac{c}{c_0} = e^{m\omega^2 r^2/2k_{\rm B}T}$$

where c_0 is the concentration at the origin. This is the equilibrium concentration distribution of macro-molecules in solution in an ultracentrifuge.

Problem 2.7 Verify that the molar constant-volume heat capacity of a perfect gas is $C_V = (3/2)R$. Compare with the law of Dulong and Petit in Eq. (2.54).

[3]Note from part (a), the velocity distribution only depends on the temperature. Hence the kinetic-energy contribution cancels in the ratio in parts (b,c).

Solution to Problem 2.7

The internal energy of one mole of a non-localized perfect gas of structureless particles at a temperature T is given in Eqs. (2.133) and (2.137)

$$E = \frac{3}{2}N_{\rm A}k_{\rm B}T = \frac{3}{2}RT \qquad ; \ R = N_{\rm A}k_{\rm B}$$

where $N_{\rm A}$ is Avagadro's number and R is the gas constant. The constant-volume heat capacity then follows as in Eq. (2.134)

$$C_V = \left(\frac{dE}{dT}\right)_{N,V} = \frac{3}{2}R$$

The law of Dulong and Petit for the same mole of structureless particles *localized* on sites is given in Eq. (2.54)

$$C_V = 3R \qquad ; \ \text{Dulong-Petit}$$

The equipartition theorem (see Prob. 2.14) implies that the additional $3R/2$ in C_V for the localized particles arises from the harmonic restoring force.

Problem 2.8 (a) Assume the transition from a classical to a quantum gas occurs at a transition temperature $T_{\rm C}$ where $n^\star \approx 1$. Show from Eq. (2.144) that this criterion can be written as

$$\frac{\hbar^2}{2m}n^{2/3} \approx \frac{1}{4\pi}k_{\rm B}T_{\rm C}$$

where $n = N/V$ is the density.

(b) Use $\rho = 0.145\,{\rm gm/cm^3}$ and $m = 6.64 \times 10^{-24}\,{\rm gm}$ for (liquid) ^{4}He. Compute $T_{\rm C}$. Compare with the measured λ-point temperature of $T_\lambda = 2.17\,^{\rm o}{\rm K}$ for the transition from the normal to the superfluid phase of ^{4}He (see [Fetter and Walecka (2003)]).

Solution to Problem 2.8

(a) The condition $n^\star \approx 1$ for the transition from a classical to a quantum gas at a transition temperature $T_{\rm C}$ is from Eq. (2.144)

$$\left(\frac{N}{V}\right)\left(\frac{h^2}{2\pi m k_{\rm B}T_C}\right)^{3/2} \approx 1$$

where $n = N/V$ is the number density. This expression is re-written as[4]

$$\frac{\hbar^2}{2m}n^{2/3} \approx \frac{1}{4\pi}k_{\rm B}T_{\rm C}$$

[4]Note the $\hbar = h/2\pi$ in this expression.

(b) Let us put in some numbers. The given values for liquid helium are

$$\rho = 0.145\,\text{gm/cm}^3 \qquad ; \; m_{\text{He}} = 6.64 \times 10^{-24}\,\text{gm}$$

This gives

$$n = \frac{\rho}{m_{\text{He}}} = 2.18 \times 10^{22}\,\text{cm}^{-3}$$

Now use the constants

$$\begin{aligned} k_{\text{B}} &= 8.620 \times 10^{-5}\,\text{eV}/^{\text{o}}\text{K} \\ \hbar^2/2m_p &= 20.74 \times 10^{-20}\,\text{eV-cm}^2 \\ m_p &= 1.673 \times 10^{-24}\,\text{gm} \end{aligned}$$

This gives

$$\begin{aligned} \frac{\hbar^2}{2m_{\text{He}}} n^{2/3} &\approx 4.08 \times 10^{-5}\,\text{eV} \\ k_{\text{B}}T_c &\approx 5.13 \times 10^{-4}\,\text{eV} \end{aligned}$$

Therefore

$$T_c \approx 5.95\,^{\text{o}}\text{K}$$

This can be compared with the measured λ-point temperature of $T_\lambda = 2.17\,^{\text{o}}\text{K}$ for the transition from the normal to the superfluid phase of liquid ^4He.

Problem 2.9 A liquid is in equilibrium with its vapor in a container at fixed (P, T). The temperature is sufficiently high, and pressure sufficiently low, that the vapor behaves as a perfect gas, Show that the chemical potential of a system in the liquid, no matter how complicated the liquid structure, is given by Eq. (2.139).

Solution to Problem 2.9

The condition for phase equilibrium at fixed (P, T) is derived from general thermodynamic arguments in Eq. (1.37)

$$\mu_1 = \mu_2 \qquad ; \text{ two phases in equilibrium}$$

The chemical potentials must be the same in the two phases. Hence if one phase behaves as a perfect gas, the chemical potential of a system in the other liquid phase, no matter how complicated the liquid structure, is given by Eq. (2.139).

Problem 2.10 Explain why the classical partition function for a perfect gas in Eq. (2.158) reproduces the result in Eq. (2.130) for all T.

Solution to Problem 2.10

The classical partition function for a perfect gas is calculated in Eq. (2.158) to be

$$(\text{p.f.})_{\text{cl}} = V\left(\frac{2\pi m k_{\text{B}}T}{h^2}\right)^{3/2}$$

This is identical to the result obtained for the quantum partition function in Eq. (2.130) under the assumption that the dimensionless quantity in Eq. (2.116) is *small*

$$\alpha^2 = \frac{h^2}{8ma^2k_{\text{B}}T} \ll 1$$

This can be achieved in several ways:

- Large mass $m \to \infty$;
- Large box $a \to \infty$;
- High temperature $T \to \infty$;
- Classical limit $h \to 0$.

It is the last limit that produces the identical results for the partition function at all temperature T.

Problem 2.11 Show that the Dirichlet integral in Eq. (2.175) gives

$$I_1(R) = 2R \qquad ; \; I_2(R) = \pi R^2 \qquad ; \; I_3(R) = \frac{4\pi}{3}R^3$$

Interpret these results.

Solution to Problem 2.11

The Dirichlet integral in Eq. (2.175) is

$$\begin{aligned} I_n(R) &\equiv \int \cdots \int dx_1 \cdots dx_n \qquad\qquad ; \; x_1^2 + x_2^2 + \cdots + x_n^2 \leq R^2 \\ &= \frac{(\sqrt{\pi}\,)^n}{\Gamma(1+n/2)}R^n \end{aligned}$$

From Eqs. (2.176)–(2.178)

$$\Gamma\left(\frac{3}{2}\right) = \frac{1}{2}\Gamma\left(\frac{1}{2}\right) = \frac{1}{2}\sqrt{\pi} \qquad ; \Gamma(2) = 1\,\Gamma(1) = 1$$
$$\Gamma\left(\frac{5}{2}\right) = \frac{3}{2}\Gamma\left(\frac{3}{2}\right) = \frac{3}{4}\sqrt{\pi}$$

This gives

$$I_1(R) = 2R \qquad ; \; I_2(R) = \pi R^2 \qquad ; I_3(R) = \frac{4\pi}{3}R^3$$

The Dirichlet integral is the "volume of an n-dimensional sphere" defined as the n-dimensional multiple integral $\int \cdots \int dx_1 \cdots dx_n$ with $x_1^2 + x_2^2 + \cdots + x_n^2 \leq R^2$. Hence the first is the length of a line, the second the area of a circle, and the third the volume of a 3-dimensional sphere.

Problem 2.12 (a) Introduce polar coordinates in two dimensions with $(x_1 = r\cos\phi,\, x_2 = r\sin\phi)$. Show $x_1^2 + x_2^2 = r^2$ and $dx_1 dx_2 = r dr d\phi$;

(b) Introduce polar-spherical coordinates in three dimensions

$$x_1 = r\sin\theta\cos\phi \qquad ; \; x_2 = r\sin\theta\sin\phi \qquad ; \; x_3 = r\cos\theta$$

Show $x_1^2 + x_2^2 + x_3^2 = r^2$ and $dx_1 dx_2 dx_3 = r^2 \sin\theta\, dr d\theta d\phi$;

(c) Introduce polar-spherical coordinates in four dimensions

$$x_1 = r\sin\beta\sin\theta\cos\phi \qquad ; \; x_2 = r\sin\beta\sin\theta\sin\phi$$
$$x_3 = r\sin\beta\cos\theta \qquad ; \; x_4 = r\cos\beta$$

Show $x_1^2 + x_2^2 + x_3^2 + x_4^2 = r^2$ and $dx_1 dx_2 dx_3 dx_4 = r^3 \sin^2\beta \sin\theta\, dr d\beta d\theta d\phi$;

(d) Do the integral over angles in each case with $0 \leq \phi \leq 2\pi$, $0 \leq \theta \leq \pi$, and $0 \leq \beta \leq \pi$. Reproduce Eq. (2.181), with C_n given by Eq. (2.184), for $n = 2, 3, 4$.

Solution to Problem 2.12

(a) We start with the transformation to polar coordinates in two dimensions

$$x_1 = r\cos\phi \qquad ; \; x_2 = r\sin\phi$$

It follows that

$$x_1^2 + x_2^2 = r^2(\cos^2\phi + \sin^2\phi) = r^2$$

The volume element transforms according to

$$dx_1dx_2 = \left|\frac{\partial(x_1,x_2)}{\partial(r,\phi)}\right| drd\phi$$

where the jacobian determinant is given by

$$\begin{aligned}\frac{\partial(x_1,x_2)}{\partial(r,\phi)} &= \begin{vmatrix}\partial x_1/\partial r & \partial x_2/\partial r \\ \partial x_1/\partial\phi & \partial x_2/\partial\phi\end{vmatrix} = \begin{vmatrix}\cos\phi & \sin\phi \\ -r\sin\phi & r\cos\phi\end{vmatrix} = r(\cos^2\phi + \sin^2\phi) \\ &= r\end{aligned}$$

Hence

$$dx_1dx_2 = r\,drd\phi$$

(b) In three dimensions we have

$$x_1 = r\sin\theta\cos\phi \qquad ;\ x_2 = r\sin\theta\sin\phi \qquad x_3 = r\cos\theta$$

Evidently

$$x_1^2 + x_2^2 + x_3^2 = r^2(\sin^2\theta + \cos^2\theta) = r^2$$

The volume element transforms according to

$$dx_1dx_2dx_3 = \left|\frac{\partial(x_1,x_2,x_3)}{\partial(r,\theta,\phi)}\right| drd\theta d\phi$$

where the jacobian determinant is now given by

$$\frac{\partial(x_1,x_2,x_3)}{\partial(r,\theta,\phi)} = \begin{vmatrix}\sin\theta\cos\phi & \sin\theta\sin\phi & \cos\theta \\ r\cos\theta\cos\phi & r\cos\theta\sin\phi & -r\sin\theta \\ -r\sin\theta\sin\phi & r\sin\theta\cos\phi & 0\end{vmatrix}$$

An expansion of the determinant in minors gives

$$\begin{aligned}\frac{\partial(x_1,x_2,x_3)}{\partial(r,\theta,\phi)} &= \cos\theta\begin{vmatrix}r\cos\theta\cos\phi & r\cos\theta\sin\phi \\ -r\sin\theta\sin\phi & r\sin\theta\cos\phi\end{vmatrix} \\ &\quad + r\sin\theta\begin{vmatrix}\sin\theta\cos\phi & \sin\theta\sin\phi \\ -r\sin\theta\sin\phi & r\sin\theta\cos\phi\end{vmatrix} \\ &= r^2(\cos^2\theta\sin\theta + \sin^3\theta) = r^2\sin\theta\end{aligned}$$

Hence the volume element becomes

$$dx_1dx_2dx_3 = r^2\sin\theta\,drd\theta d\phi$$

(c) In four dimensions the transformation to polar coordinates is

$$x_1 = r\sin\beta\sin\theta\cos\phi \qquad ; \; x_2 = r\sin\beta\sin\theta\sin\phi$$
$$x_3 = r\sin\beta\cos\theta \qquad ; \; x_4 = r\cos\beta$$

Here

$$x_1^2 + x_2^2 + x_3^2 + x_4^2 = r^2\sin^2\beta(\sin^2\theta + \cos^2\theta) + r^2\cos^2\beta = r^2$$

The jacobian determinant is

$$\frac{\partial(x_1, x_2, x_3, x_4)}{\partial(r, \beta, \theta, \phi)} = \begin{vmatrix} \sin\beta\sin\theta\cos\phi & \sin\beta\sin\theta\sin\phi & \sin\beta\cos\theta & \cos\beta \\ r\cos\beta\sin\theta\cos\phi & r\cos\beta\sin\theta\sin\phi & r\cos\beta\cos\theta & -r\sin\beta \\ r\sin\beta\cos\theta\cos\phi & r\sin\beta\cos\theta\sin\phi & -r\sin\beta\sin\theta & 0 \\ -r\sin\beta\sin\theta\sin\phi & r\sin\beta\sin\theta\cos\phi & 0 & 0 \end{vmatrix}$$

An expansion in minors then gives

$$\begin{aligned} -\frac{\partial(x_1, x_2, x_3, x_4)}{\partial(r, \beta, \theta, \phi)} &= r\cos^2\beta\sin^2\beta\frac{\partial(x_1, x_2, x_3)}{\partial(r, \theta, \phi)} + r\sin^4\beta\frac{\partial(x_1, x_2, x_3)}{\partial(r, \theta, \phi)} \\ &= r\sin^2\beta\frac{\partial(x_1, x_2, x_3)}{\partial(r, \theta, \phi)} \end{aligned}$$

Where the lower-dimension jacobian is just the one calculated in part (b). Therefore

$$dx_1dx_2dx_3dx_4 = r^3\sin^2\beta\sin\theta\, drd\beta d\theta d\phi$$

(d) Let us do the integrals over angles in each of these cases

$$\int_0^{2\pi} d\phi = 2\pi$$
$$\int_0^{\pi} \sin\theta\, d\theta \int_0^{2\pi} d\phi = -2\pi\left[\cos\theta\right]_0^{\pi} = 4\pi$$
$$\int_0^{\pi} \sin^2\beta\, d\beta \int_0^{\pi} \sin\theta\, d\theta \int_0^{2\pi} d\phi = 4\pi\left[\frac{\beta}{2} - \frac{1}{4}\sin 2\beta\right]_0^{\pi} = 2\pi^2$$

The result given in Eqs. (2.181) and (2.184) is

$$\int\cdots\int dx_1\cdots dx_n = nC_n r^{n-1}dr \qquad ; \text{ over all angles}$$
$$C_n = \frac{(\sqrt{\pi}\,)^n}{\Gamma(1+n/2)}$$

This agrees with the above for $n = 2, 3, 4$.

Problem 2.13 (a) The energy of the one-dimensional simple harmonic oscillator is given in Eq. (2.159). Make a phase-space plot where the ordinate is p and the abscissa is x. Show the constant-energy phase-space orbit is an ellipse with semi-major axis $a = (2\varepsilon/m\omega^2)^{1/2}$ and semi-minor axis $b = (2m\varepsilon)^{1/2}$;

(b) Verify that the area of the ellipse is $\pi ab = 2\pi\varepsilon/\omega$;

(c) In quantum mechanics, the energy is quantized with $\varepsilon_n = \hbar\omega(n + 1/2)$. Show the area between the states with $n+1$ and n is exactly that given in Eq. (2.153).

Solution to Problem 2.13

(a) The energy of the one-dimensional simple harmonic oscillator $\varepsilon(p, x)$ is given in Eq. (2.159)

$$\varepsilon = \frac{p^2}{2m} + \frac{1}{2}m\omega^2 x^2$$

The constant-energy phase-space orbit is then an *ellipse*

$$\left(\frac{x}{a}\right)^2 + \left(\frac{p}{b}\right)^2 = 1$$

with semi-major axis a, and semi-minor axis b, given respectively by

$$a = \sqrt{\frac{2\varepsilon}{m\omega^2}} \qquad ; \; b = \sqrt{2m\varepsilon}$$

See Fig. 2.2.

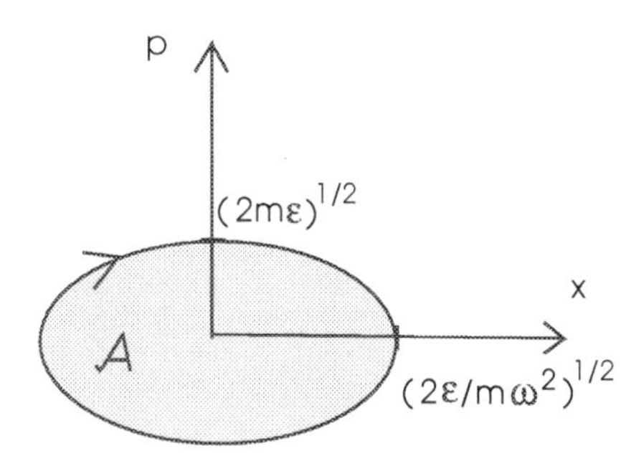

Fig. 2.2 Constant-energy phase-space orbit of the simple harmonic oscillator. This is an ellipse with area $\mathcal{A} = \pi ab = 2\pi\varepsilon/\omega$.

(b) The area of an ellipse is given by πab. Thus the area of the phase-space orbit is

$$\mathcal{A} = \pi ab = \frac{2\pi\varepsilon}{\omega}$$

(c) In quantum mechanics, the energy of the one-dimensional simple harmonic oscillator is quantized according to

$$\varepsilon_n = \hbar\omega(n + 1/2) \qquad ; \ n = 0, 1, 2, \cdots$$

This implies that the area of the phase-space orbit is quantized according to

$$\begin{aligned} \mathcal{A}_n &= \frac{2\pi}{\omega}\hbar\omega(n + 1/2) \qquad ; \ n = 0, 1, 2, \cdots \\ &= h(n + 1/2) \end{aligned}$$

The phase-space area *between* orbits is then

$$\mathcal{A}_{n+1} - \mathcal{A}_n = h$$

This reproduces the result given in Eq. (2.153).

Problem 2.14 (a) Prove that if the classical expression for the energy of a system can be written as

$$\varepsilon(q_1, \cdots, q_n; p_1, \cdots, p_n) = \sum_{i=1}^{n} a_i q_i^2 + \sum_{i=1}^{n} b_i p_i^2$$

where all the a's and b's are non-zero and positive, then the classical internal energy of the assembly is $nk_{\text{B}}T$ per system.

(b) Show that if only l of the a's and m of the b's are different from zero, the classical value of the internal energy is $[(l+m)/2]k_{\text{B}}T$ per system.

This is one statement of the *classical equipartition theorem.*

Solution to Problem 2.14

(a) From Eqs. (2.44) and (2.155), the mean value of the energy in classical statistical mechanics for a system moving in $2n$-dimensional phase space is

$$\frac{E}{N} = \frac{(1/h^n)\int\cdots\int dp_1\cdots dp_n\, dq_1\cdots dq_n\, \varepsilon\, e^{-\varepsilon/k_{\text{B}}T}}{(1/h^n)\int\cdots\int dp_1\cdots dp_n\, dq_1\cdots dq_n\, e^{-\varepsilon/k_{\text{B}}T}}$$

It is given that the energy is a positive quadratic form in the generalized coordinates and momenta

$$\varepsilon(q_1,\cdots,q_n;p_1,\cdots,p_n) = \sum_{i=1}^{n} a_i q_i^2 + \sum_{i=1}^{n} b_i p_i^2$$

The exponentials all *factor* in the numerator and denominator, and the above expression becomes

$$\frac{E}{N} = \sum_{i=1}^{n} I(q_i) + \sum_{i=1}^{n} I(p_i)$$

where all the contributions take the same form

$$I(q) = \frac{\int_{-\infty}^{\infty} dq\, aq^2\, e^{-aq^2/k_\mathrm{B}T}}{\int_{-\infty}^{\infty} dq\, e^{-aq^2/k_\mathrm{B}T}} = k_\mathrm{B}T\, \frac{\int_{-\infty}^{\infty} dx\, x^2\, e^{-x^2}}{\int_{-\infty}^{\infty} dx\, e^{-x^2}}$$

Notice that

- The integrals over all the coordinates not involved in the ε in the integrand in the numerator of the expression for E/N *factor and cancel in the ratio*;
- $I(q)$ is independent of a, and an identical result is obtained for $I(p)$, independent of b

$$I(p) = \frac{\int_{-\infty}^{\infty} dp\, bp^2\, e^{-bp^2/k_\mathrm{B}T}}{\int_{-\infty}^{\infty} dp\, e^{-bp^2/k_\mathrm{B}T}} = k_\mathrm{B}T\, \frac{\int_{-\infty}^{\infty} dx\, x^2\, e^{-x^2}}{\int_{-\infty}^{\infty} dx\, e^{-x^2}}$$

Now from Eq. (2.127)

$$\int_{-\infty}^{\infty} e^{-\lambda t^2}\, dt = \left(\frac{\pi}{\lambda}\right)^{1/2}$$

$$\int_{-\infty}^{\infty} t^2\, e^{-\lambda t^2}\, dt = \frac{1}{2\lambda}\left(\frac{\pi}{\lambda}\right)^{1/2}$$

where the second relation follows from differentiation w.r.t. λ. Hence, upon setting $\lambda = 1$,

$$I(q) = I(p) = \frac{1}{2}k_\mathrm{B}T$$

It follows that

$$\frac{E}{N} = \frac{n}{2}k_\mathrm{B}T + \frac{n}{2}k_\mathrm{B}T = nk_\mathrm{B}T$$

(b) Suppose only l of the a's and m of the b's are different from zero so that

$$\varepsilon(q_1, \cdots, q_n; p_1, \cdots, p_n) = \sum_{i=1}^{l} a_i q_i^2 + \sum_{i=1}^{m} b_i p_i^2$$

Then the above gives

$$\frac{E}{N} = \sum_{i=1}^{l} I(q_i) + \sum_{i=1}^{m} I(p_i)$$

$$= \frac{l}{2} k_B T + \frac{m}{2} k_B T = \frac{(l+m)}{2} k_B T$$

This problem provides a demonstration of the *equipartition theorem* of classical statistical mechanics; there is a contribution of $k_B T/2$ to the internal energy for every quadratic term in the hamiltonian.[5]

Problem 2.15 Suppose, in contrast to statistical assumption I in Eq. (1.41), that the summand in Eq. (2.2) were to be weighted with some probability $P(m)$, where $0 \leq P(m) \leq 1$, with $P(N/2) = 1$ and $P'(N/2) = 0$.[6] Show the entropy is still given by Eq. (2.12), and the argument in Eqs. (2.13)–(2.14) is unchanged. Discuss the implications of this observation.

Solution to Problem 2.15

The first of our two basic statistical assumptions in Eq. (1.41) reads

All complexions consistent with (E, V, N) are a priori equally probable.

In the two-level problem analyzed in section 2.1, this lead to the total number of complexions in Eq. (2.4)

$$\Omega = \sum_{m=0}^{N} \frac{N!}{m!(N-m)!} = 2^N$$

The entropy then followed from our second statistical assumption in Eq. (1.46)

$$S = k_B \ln \Omega = N k_B \ln 2$$

[5]Here the hamiltonian $h(q, p) = \varepsilon(q_1, \cdots, q_n; p_1, \cdots, p_n)$.

[6]The last two arguments are corrected here to $m/N = 1/2$.

It was shown in section 2.1 that for very large N, one can reproduce the above result for the entropy by retaining just the *largest term in the sum for* Ω. To find that term, one employs Stirling's formula and sets the derivative of the summand equal to zero. This yields

$$\frac{m}{N} = \frac{1}{2}$$
$$\Omega \doteq \frac{N!}{(N/2)!(N-N/2)!}$$

Then, from Eq. (2.11),

$$S = k_{\rm B} \ln \Omega = N k_{\rm B} \ln 2$$

Now suppose, in contrast to statistical assumption I in Eq. (1.41), that the summand in Eq. (2.2) were to be weighted with some probability $P(m)$, where $0 \leq P(m) \leq 1$, with $P(N/2) = 1$ and $P'(N/2) = 0$. Then

$$\tilde{\Omega} = \sum_{m=0}^{N} P(m) \frac{N!}{m!(N-m)!}$$

The argument in Eqs. (2.13)–(2.14) implies that we can still replace the summand by the largest term, obtained by setting the derivative of the summand equal to zero. Under the stated conditions, the previous result is unchanged

$$\frac{m}{N} = \frac{1}{2}$$
$$\tilde{\Omega} \doteq \frac{N!}{(N/2)!(N-N/2)!}$$

and the entropy remins the same

$$S = k_{\rm B} \ln \tilde{\Omega} = N k_{\rm B} \ln 2$$

The implication is that our first statistical assumption is much stronger than necessary, and all that is required is to have assumption I hold in the close vicinity of the largest term in the sum over complexions.[7]

[7]An interesting and thought-provoking result!

Chapter 3

Applications of the Microcanonical Ensemble

Problem 3.1 Show that the second of Eqs. (3.4) follows from the first.

Solution to Problem 3.1

If the internal energies are additive, then the internal partition function factors [see Eqs. (3.1)–(3.3)]

$$(\text{p.f.}) = (\text{p.f.})^{(1)}(\text{p.f.})^{(2)}(\text{p.f.})^{(3)}\cdots \qquad ;\ \text{factors}$$

The Helmholtz free energy is related to the partition function in Eqs. (2.103)

$$A = -Nk_{\text{B}}T\ln(\text{p.f.}) \qquad ;\ \text{Helmholtz free energy}$$

Thus one immediately arrives at the first of Eqs. (3.4)

$$A = A^{(1)} + A^{(2)} + A^{(3)} + \cdots \qquad ;\ \text{additive}$$

which states that the internal free energies are also additive.

The entropy is related to the Helmholtz free energy by

$$S = \frac{E-A}{T}$$

Since the energies are additive, as are the free energies, the internal entropies are also additive

$$S = S^{(1)} + S^{(2)} + S^{(3)} + \cdots$$

This is the second of Eqs. (3.4).

Problem 3.2 Use the analysis in Eqs. (2.119)–(2.124) to convert the sum to an integral, and verify Eq. (3.39).

Solution to Problem 3.2

The rotational partition function in Eqs. (3.37)–(3.38) is

$$(\text{p.f.})_{\text{rot}} = \sum_{l=0}^{\infty}(2l+1)e^{-l(l+1)\theta_{\text{R}}/T} \qquad ; \text{ rotation}$$

We follow the analysis in Eqs. (2.119)–(2.124). Write this as

$$(\text{p.f.})_{\text{rot}} = \sum_{l=0}^{\infty}(2l+1)\Delta l \; e^{-l(l+1)\theta_{\text{R}}/T} \qquad ; \; \Delta l = 1$$

Define

$$x_l \equiv \frac{l(l+1)}{T}$$
$$\Delta x_l \equiv \frac{1}{2}[(x_{l+1} - x_l) + (x_l - x_{l-1})] = \frac{(2l+1)\Delta l}{T}$$

Then

$$(\text{p.f.})_{\text{rot}} = T\sum_{l=0}^{\infty}\Delta x_l \; f(x_l) \qquad ; \; f(x_l) = e^{-x_l\theta_{\text{R}}}$$

Here we have associated the mean value $f(x_l)$ with the interval Δx_l. In the limit $T \to \infty$, the quantity Δx_l becomes a differential dx, and this expression becomes an integral

$$(\text{p.f.})_{\text{rot}} \to T\int_0^{\infty} dx\, f(x) \qquad ; \; T \to \infty$$

Hence we obtain

$$(\text{p.f.})_{\text{rot}} \to T\int_0^{\infty} dx\, e^{-x\theta_{\text{R}}} \qquad ; \; T \to \infty$$
$$= \int_0^{\infty} du\, e^{-u\theta_{\text{R}}/T}$$

This is the result in Eq. (3.39).

Problem 3.3 Use Eq. (3.43) to show that the electronic partition function in Eq. (3.41) contributes only a constant to the energy, and thus makes a vanishing contribution to the constant-volume heat capacity.

Solution to Problem 3.3

The partition function for the unexcited electronic system is given in Eq. (3.41)

$$(\text{p.f.})_{\text{el}} = e^{-\varepsilon_0/k_{\text{B}}T} \qquad ; \text{ electronic ground state}$$

This makes an *additive* contribution to the Helmholtz free energy in Eq. (3.42)

$$\begin{aligned} A(T,V,N)_{\text{el}} &= -Nk_{\text{B}}T \ln{(\text{p.f.})_{\text{el}}} \\ &= N\varepsilon_0 \end{aligned}$$

The corresponding electronic energy is obtained from this Helmholtz free energy through Eq. (3.43)

$$\begin{aligned} E_{\text{el}} &= -T^2 \frac{\partial}{\partial T}\left(\frac{A_{\text{el}}}{T}\right)_{N,V} \qquad ; \text{ energy} \\ &= N\varepsilon_0 \end{aligned}$$

and we recover an obvious result. Since this additional energy is independent of temperature, there is no contribution to the constant-volume heat capacity in Eq. (3.44)

$$\begin{aligned} (C_V)_{\text{el}} &= \left(\frac{\partial E_{\text{el}}}{\partial T}\right)_{N,V} \qquad ; \text{ heat capacity} \\ &= 0 \end{aligned}$$

Problem 3.4 Show from Eqs. (3.43)–(3.44) that every mode in the internal partition function that has $(\text{p.f.})_{\text{mode}} \propto T^{\nu}$ in the high-temperature limit, will contribute νR to the molar constant-volume heat capacity.

Solution to Problem 3.4

The additive contribution of this internal partition function to the Helmholtz free energy is given in Eq. (3.42)

$$A(T,V,N)_{\text{mode}} = -k_{\text{B}}T \ln{(\text{p.f.})^N_{\text{mode}}}$$

If $(\text{p.f.})_{\text{mode}} \propto T^{\nu}$, then the temperature dependence of this expression arises from an additive contribution

$$A(T,V,N)_{\text{mode}} = -\nu Nk_{\text{B}}T \ln T$$

The corresponding contribution to the internal energy follows from Eq. (3.43)

$$\begin{aligned} E_{\text{mode}} &= -T^2 \frac{\partial}{\partial T}\left(\frac{A_{\text{mode}}}{T}\right)_{N,V} \\ &= \nu N k_{\text{B}} T \end{aligned}$$

The additive contribution to the molar constant-volume heat capacity then follows from Eq. (3.44)

$$\begin{aligned} (C_V)_{\text{mode}} &= \left(\frac{\partial E_{\text{mode}}}{\partial T}\right)_{N,V} \\ &= \nu N_{\text{A}} k_{\text{B}} \\ &= \nu R \end{aligned}$$

which is the stated answer.

Problem 3.5 Pick representative values of $(\theta_{\text{R}}, \theta_{\text{V}})$, and make a good numerical calculation of the molar heat capacity of a diatomic gas $\mathcal{C}_V/R$ in Fig. 3.4 in the text.[1]

Solution to Problem 3.5

From Eqs. (3.42)–(3.44)

$$\begin{aligned} A(T, V, N)_{\text{rot-vib}} &= -k_{\text{B}} T \left[\ln{(\text{p.f.})_{\text{vib}}^N} + \ln{(\text{p.f.})_{\text{rot}}^N}\right] \\ E &= -T^2 \frac{\partial}{\partial T}\left(\frac{A}{T}\right)_{N,V} \\ C_V &= \left(\frac{\partial E}{\partial T}\right)_{N,V} \end{aligned}$$

From Eqs. (3.37)–(3.38) and (3.34)

$$\begin{aligned} (\text{p.f.})_{\text{rot}} &= \sum_{l=0}^{\infty} (2l+1) e^{-\theta_{\text{R}} l(l+1)/T} \\ (\text{p.f.})_{\text{vib}} &= \frac{e^{-\theta_{\text{V}}/2T}}{1 - e^{-\theta_{\text{V}}/T}} \end{aligned}$$

Then

$$\mathcal{C}_V = (\mathcal{C}_V)_{\text{rot}} + (\mathcal{C}_V)_{\text{vib}}$$

[1] Recall the arguments in Eqs. (2.51)–(2.53).

From Eqs. (2.51)–(2.53), the vibrational contribution to the molar heat capacity in units of R from a one-dimensional simple harmonic oscillator is

$$(\mathcal{C}_V)_{\text{vib}} = \left(\frac{\theta_{\text{V}}}{T}\right)^2 \frac{e^{\theta_{\text{V}}/T}}{(e^{\theta_{\text{V}}/T} - 1)^2}$$

The rotational energy is calculated from the above as

$$E_{\text{rot}} = \frac{Nk_{\text{B}}}{(\text{p.f.})_{\text{rot}}} \sum_{l=0}^{\infty} [\theta_{\text{R}} l(l+1)](2l+1)\, e^{-\theta_{\text{R}} l(l+1)/T}$$

The rotational contribution to the molar heat capacity in units of R is then calculated to be

$$(\mathcal{C}_V)_{\text{rot}} = \left(\frac{\theta_{\text{R}}}{T}\right)^2 \left[\langle\, l^2(l+1)^2 \rangle - \langle\, l(l+1)\rangle^2\right]$$

where the expectation values are given by

$$\langle\, F(l)\rangle = \frac{1}{(\text{p.f.})_{\text{rot}}} \sum_{l=0}^{\infty} F(l)\,(2l+1)\, e^{-\theta_{\text{R}} l(l+1)/T}$$

The resulting molar rotation and vibration constant-volume heat capacity is plotted in Fig. 3.1 for $\theta_{\text{R}} = 10\,^{\text{o}}\text{K}$ and $\theta_{\text{V}} = 100\,^{\text{o}}\text{K}$.

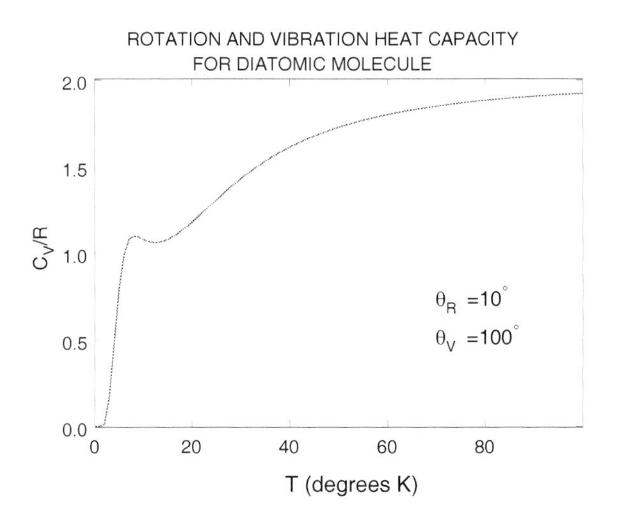

Fig. 3.1 Sum of the rotation and vibration contributions to the molar heat capacity of a diatomic molecule vs. the temperature in ^{o}K. Here $\theta_{\text{R}} = 10\,^{\text{o}}\text{K}$ and $\theta_{\text{V}} = 100\,^{\text{o}}\text{K}$. (Compare Fig. 3.4 in the text.)

Problem 3.6 (a) Make a good numerical calculation of the molar heat capacity of molecular H_2 obtained from Eq. (3.52) with $A = 1$ and $\rho = 2$. Express the result as a function of T/θ_R;

(b) Now repeat the calculation for a metastable assembly that retains the high-temperature ratio of $N_{\text{ortho-H}_2}/N_{\text{para-H}_2} = 3/1$ down to low temperatures [see Eq. (3.58)];

(c) Compare these results with each other and with experiment as sketched in Fig. 3.7 in the text.

Solution to Problem 3.6

(a) In the case of statistical equilibrium, we have for $A = 1$ and $\rho = 2$ from Eq. (3.52)

$$(\text{p.f.})_{\text{rot}} = \sum_{\text{even } l} (2l+1)e^{-l(l+1)\theta_R/T} + 3\sum_{\text{odd } l} (2l+1)e^{-l(l+1)\theta_R/T}$$

From Prob. 3.5

$$(C_V)_{\text{rot}} = \left(\frac{\theta_R}{T}\right)^2 \left[\langle l^2(l+1)^2\rangle - \langle l(l+1)\rangle^2\right]$$

where the expectation values are given by

$$\langle F(l)\rangle = \frac{1}{(\text{p.f.})_{\text{rot}}} \times \left[\sum_{\text{even } l} F(l)(2l+1)e^{-l(l+1)\theta_R/T} + 3\sum_{\text{odd } l} F(l)(2l+1)e^{-l(l+1)\theta_R/T}\right]$$

(b) With the high-T distribution frozen in, we have from Eq. (3.58)

$$C_V = \frac{3}{4}C_V^{\text{ortho-H}_2} + \frac{1}{4}C_V^{\text{para-H}_2}$$

where

$$(\text{p.f.})^{\text{para}} = \sum_{\text{even } l} (2l+1)e^{-l(l+1)\theta_R/T}$$
$$(\text{p.f.})^{\text{ortho}} = 3\sum_{\text{odd } l} (2l+1)e^{-l(l+1)\theta_R/T}$$

The individual heat capacities in this case are

$$C_V = \left(\frac{\theta_R}{T}\right)^2 \left[\langle l^2(l+1)^2\rangle - \langle l(l+1)\rangle^2\right]$$

where the expectation values are given by

$$\langle F(l) \rangle^{\text{para}} = \frac{1}{(\text{p.f.})^{\text{para}}} \sum_{\text{even } l} F(l)(2l+1)e^{-l(l+1)\theta_{\text{R}}/T}$$

$$\langle F(l) \rangle^{\text{ortho}} = \frac{3}{(\text{p.f.})^{\text{ortho}}} \sum_{\text{odd } l} F(l)(2l+1)e^{-l(l+1)\theta_{\text{R}}/T}$$

The various results for the molar heat capacity of H_2 are plotted in Fig. 3.2. Three curves are shown: a statistical mixture of ortho- and para-hydrogen; pure para-hydrogen; and pure ortho-hydrogen. If the high-temperature statistical ratio is "frozen in", then one has $\mathcal{C}_V = \mathcal{C}_V^{\text{para}}/4 + 3\,\mathcal{C}_V^{\text{ortho}}/4$, and this reproduces the data (see Fig. 3.7 in the text.)

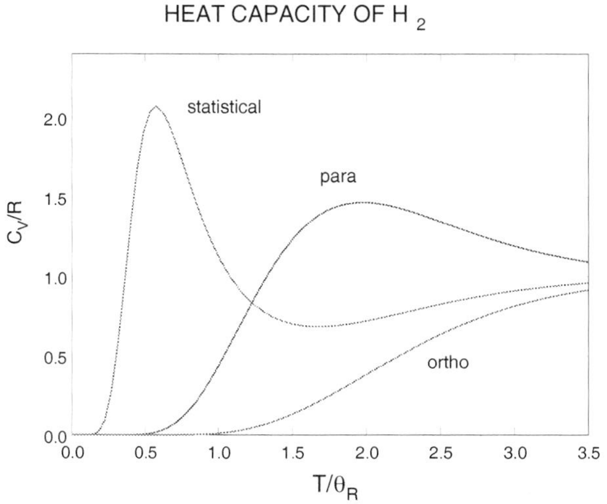

Fig. 3.2 Molar heat capacity of H_2 as a function of T/θ_{R}. Three curves are shown: a statistical mixture of ortho- and para-hydrogen [See Eq. (3.52)]; pure para-hydrogen; and pure ortho-hydrogen. If the high-temperature statistical ratio is "frozen in", then one has $\mathcal{C}_V = \mathcal{C}_V^{\text{para}}/4 + 3\,\mathcal{C}_V^{\text{ortho}}/4$; this reproduces the data (see Fig. 3.7 in the text.)

Problem 3.7 Verify the high-temperature relations satisfied by the rotational partition function in Eqs. (3.54).

Solution to Problem 3.7

We are asked to show that in the high-temperature limit, the sum over even l and sum over odd l give one-half the sum over all l for the rotational partition function for the diatomic molecule

$$\sum_{\text{even } l}(2l+1)e^{-l(l+1)\theta_{\rm R}/T} = \sum_{\text{odd } l}(2l+1)e^{-l(l+1)\theta_{\rm R}/T} \qquad ; \; T\to\infty$$
$$= \frac{1}{2}\sum_{l}(2l+1)e^{-l(l+1)\theta_{\rm R}/T}$$

Consider the first sum, and repeat the argument in the solution to Prob. 3.2

$$\sum_{\text{even } l}(2l+1)e^{-l(l+1)\theta_{\rm R}/T} = \frac{1}{2}\sum_{\text{even } l}^{\infty}(2l+1)\Delta l\; e^{-l(l+1)\theta_{\rm R}/T} \qquad ; \; \Delta l = 2$$

Note that here $\Delta l = 2$. Define

$$x_l \equiv \frac{l(l+1)}{T}$$
$$\Delta x_l \equiv x_{l+1} - x_{l-1} = \frac{(2l+1)\Delta l}{T}$$

Then

$$\sum_{\text{even } l}(2l+1)e^{-l(l+1)\theta_{\rm R}/T} = \frac{1}{2}T\sum_{\text{even } l}\Delta x_l\; e^{-x_l\theta_{\rm R}}$$

Here we have again associated the mean value $e^{-x_l\theta_{\rm R}}$ with the interval Δx_l. In the limit $T\to\infty$, the quantity Δx_l becomes a differential dx, and this expression becomes an integral. Hence we obtain

$$(\text{p.f.})_{\rm rot} \to \frac{1}{2}T\int_0^\infty dx\, e^{-x\theta_{\rm R}} \qquad ; \; T\to\infty$$
$$= \frac{1}{2}\int_0^\infty du\, e^{-u\theta_{\rm R}/T}$$

The same analysis holds for the sum over odd l. Hence we arrive at the result in Eqs. (3.54)

$$\sum_{\text{even } l} = \sum_{\text{odd } l} = \frac{1}{2}\sum_{l} \qquad ; \; T\to\infty$$

Problem 3.8 The high-temperature limit of the rotational partition function for the homonuclear diatomic molecule in Eq. (3.55) contains a factor of ρ^2 for the nuclear spin degeneracy, where $\rho = 2\mathcal{I}+1$. Work through the corresponding Helmholtz free energy.

(a) What is the effect of this degeneracy factor on the energy E? On the heat capacity C_V?

(b) What is the effect of this factor on the entropy S? On the chemical potential μ?

Solution to Problem 3.8

The high-temperature limit of the rotational partition function for the homonuclear diatomic molecule in Eq. (3.55) is

$$(\text{p.f.})_{\text{rot}} = \frac{1}{2}\left[\rho^2 \sum_l (2l+1)e^{-l(l+1)\theta_{\text{R}}/T}\right] \qquad ; T \to \infty$$

The corresponding Helmholtz free energy in Eq. (3.42) receives an *additive* contribution from the factor of ρ^2 of

$$A_\rho(T,V,N) = -Nk_{\text{B}}T \ln \rho^2$$

(a) From Eqs. (3.43)–(3.44), the contribution of this term to the internal energy and heat capacity *vanishes*

$$E_\rho = -T^2 \frac{\partial}{\partial T}\left(\frac{A_\rho}{T}\right)_{N,V} = 0$$

$$(C_V)_\rho = \left(\frac{\partial E_\rho}{\partial T}\right)_{N,V} = 0$$

(b) The ρ^2 term merely produces an increase in the number of complexions. There will be a corresponding increase in the entropy obtained from Eq. (1.26) as

$$S_\rho = -\left(\frac{\partial A_\rho}{\partial T}\right)_{N,V} = Nk_{\text{B}} \ln \rho^2$$

The corresponding decrease in chemical potential obtained from Eq. (1.26) is

$$\mu_\rho = \left(\frac{\partial A_\rho}{\partial N}\right)_{T,V} = -k_{\text{B}}T \ln \rho^2$$

Problem 3.9 (a) Suppose the vibrational state changes, while the electronic state remains unchanged. Derive the first selection rule in Eqs. (3.63) for the operator in Eq. (3.62) taken between simple-harmonic-oscillator vibrational eigenstates $\psi_n^{\text{vib}}(x)$ with the energy eigenvalues given in Eq. (3.33);

(b) Use the properties of the matrix element of the spherical harmonic in Eq. (3.62) taken between the rotational eigenstates in Eq. (3.36) to show that $\Delta l = 0, \pm 1$;

(c) Use the parity of the operator $\mathbf{r}$ to show that l must change, and hence arrive at the second selection rule in Eqs. (3.63).[2]

Solution to Problem 3.9

(a) The wave function for the diatomic molecule is given in Eq. (3.46)

$$\Psi_{\text{tot}} = \psi_{\text{el}}(\mathbf{r}_i; \mathbf{r})\, \psi_{\text{vib}}(x)\, \psi_{\text{rot}}(\theta, \phi)\, \psi_{\text{nuclear spins}}$$

where $\mathbf{r}$ is the relative nuclear coordinate. If the electronic state is unchanged, then the electronic dipole matrix element yields the electric dipole operator for the molecule (see Prob. 3.10)

$$e_p \langle \Psi_{\text{el}} | \mathbf{d} | \Psi_{\text{el}} \rangle = e_p \zeta \mathbf{r}$$

Here we investigate the matrix element of the remaining electric dipole operator $\mathbf{r}$ between the vibration-rotation states

$$M_{fi} \propto \langle \psi_f^{\text{vib}}\, \psi_f^{\text{rot}}\, |\mathbf{r}|\, \psi_i^{\text{vib}}\, \psi_i^{\text{rot}} \rangle$$

In spherical coordinates, the components of the vector $\mathbf{r}$ can be written

$$r_{1m} = (r_0 + x) \left(\frac{4\pi}{3}\right)^{1/2} Y_{1m}(\theta, \phi) \qquad ; \; m = 0, \pm 1$$

The result for the vibrations is well-known. The only non-zero matrix elements of x for the simple harmonic oscillator are (see [Schiff (1968)])

$$\langle n+1 | x | n \rangle = \left(\frac{\hbar}{2\mu\omega}\right)^{1/2} \sqrt{n+1}$$

$$\langle n-1 | x | n \rangle = \left(\frac{\hbar}{2\mu\omega}\right)^{1/2} \sqrt{n}$$

This leads to the first selection rule in Eqs. (3.63).[3]

---

[2] [Schiff (1968)] and [Edmonds (1974)] are good resources for this problem.

[3] The selection rules in Eqs. (3.63) apply to the case where the vibrational state changes. I see no reason why there cannot be low-frequency transitions, with $\Delta n = 0$, arising from the r_0 term in r_{1m}.

(b) The matrix element of the spherical harmonic between spherical harmonics is given in [Edmonds (1974)]

$$\langle l_1, m_1 | Y_{l_2,m_2}(\theta,\phi) | l_3, m_3 \rangle = (-1)^{m_1} \left[\frac{(2l_1+1)(2l_2+1)(2l_3+1)}{4\pi} \right]^{1/2}$$
$$\times \begin{pmatrix} l_1 & l_2 & l_3 \\ 0 & 0 & 0 \end{pmatrix} \begin{pmatrix} l_1 & l_2 & l_3 \\ -m_1 & m_2 & m_3 \end{pmatrix}$$

where the last two factor are 3-j symbols (C-G coefficients). They both vanish unless

$$|l_1 - l_3| \leq l_2 \leq l_1 + l_3$$

Since $l_2 = 1$, this implies $\Delta l = 0, \pm 1$.

(c) The first 3-j symbol above vanishes unless $l_1+l_2+l_3$ is an even integer (parity); hence, $\Delta l = 0$ is ruled out. This leads to the second selection rule in Eqs. (3.63).

Problem 3.10 The electric dipole moment for a collection of charges is defined by $|e|\mathbf{d} = \sum_p q_p \mathbf{r}_p$ where the sum goes over all the charges. The electric dipole moment for the diatomic molecule in Fig. 3.1 in the text is therefore

$$\mathbf{d} = Z_A \mathbf{r}_A + Z_B \mathbf{r}_B - \sum_i \mathbf{r}_i$$

(a) Show that the electric dipole moment of two neutral, spherically symmetric atoms placed at $\mathbf{r}_A$ and $\mathbf{r}_B$ vanishes;

(b) The ground-state electronic wave function in the molecule is of the form $\Psi_{\rm el}(\mathbf{r}_i; \mathbf{r})$. Define the expectation value of the last term in $\mathbf{d}$ by

$$\langle \Psi_{\rm el} | \sum_i \mathbf{r}_i | \Psi_{\rm el} \rangle \equiv Z_A \mathbf{r}_A + Z_B \mathbf{r}_B - \zeta \mathbf{r}$$

Hence conclude that the effective dipole moment arises from the redistribution of electronic charge in the molecule[4]

$$|e|\mathbf{d} = |e|\zeta \mathbf{r}$$

[4]There is a theorem that either parity or time-reversal invariance implies the vanishing of an electric dipole moment; however, here there is an external vector $\mathbf{r}$ in the electron problem arising from the fact that the two nuclei are heavy and fixed. Indeed, diatomic molecules do exhibit electric dipole moments.

(c) What is the electric dipole moment of a homonuclear diatomic molecule? Give your argument.

Solution to Problem 3.10

(a) The dipole operator $e_p\mathbf{d}$ for the diatomic molecule in Fig. 3.1 in the text is

$$\mathbf{d} = Z_A\mathbf{r}_A + Z_B\mathbf{r}_B - \sum_i \mathbf{r}_i$$

If each of the two atoms is completely localized, separated, and unaffected by the presence of the other, then the electron contribution to the dipole moment of the molecule will be

$$-\langle\Psi_{\rm el}|\sum_i \mathbf{r}_i\,|\Psi_{\rm el}\rangle \equiv -Z_A\mathbf{r}_A - Z_B\mathbf{r}_B$$

The magnitude of the electronic contribution for the neutral, spherically symmetric atom will be precisely that of the nuclear charge, and the only remaining vector for the atom will be the displacement of the nucleus from the center-of-mass. Hence, in this case, the dipole moment of the molecule vanishes

$$\langle\Psi_{\rm el}|\sum_p q_p\mathbf{r}_p|\Psi_{\rm el}\rangle = 0$$

(b) Now let the atoms come together to form the diatomic molecule. The electron clouds will be distorted by

- The shape of the new attractive nuclear Coulomb potential;
- The electron Coulomb repulsion and the shape of the new, self-consistent, electron electrostatic potential;
- The Pauli principle repulsion between the like fermions (electrons).[5]

The electron contribution to the expectation value of the dipole operator $e_p\mathbf{d}$ in the *actual* molecule will therefore take the form

$$-\langle\Psi_{\rm el}|\sum_i \mathbf{r}_i\,|\Psi_{\rm el}\rangle \equiv -Z_A\mathbf{r}_A - Z_B\mathbf{r}_B + \zeta\mathbf{r}$$

[5] A properly antisymmetrized wave function $\psi(\mathbf{r}_1,\mathbf{r}_2) = (1/\sqrt{2})\,[\psi_1(\mathbf{r}_1)\psi_2(\mathbf{r}_2) - \psi_1(\mathbf{r}_2)\psi_2(\mathbf{r}_1)]$ vanishes for $\mathbf{r}_2 \to \mathbf{r}_1$.

The final contribution must be proportional to $\mathbf{r}$, which is the only remaining vector in the problem. In this case

$$\langle\Psi_{\rm el}|\sum_p q_p\mathbf{r}_p|\Psi_{\rm el}\rangle = e_p\zeta\mathbf{r}$$

(c) For a homonuclear diatomic molecule, this expression must vanish. By symmetry, there is no direction in which this dipole moment can point.[6]

Problem 3.11 A more accurate treatment of the internal motion of a diatomic molecule is obtained by writing $r = r_0 + x$ in the second term on the r.h.s. of Eqs. (3.25), and then expanding in x. Discuss the effects of the additional *rotation-vibration coupling*.

Solution to Problem 3.11

With the indicated expansion, the second term in Eq. (3.25) becomes

$$\frac{1}{2\mu(r_0+x)^2}\mathbf{L}^2 = \frac{1}{2\mu r_0^2}\mathbf{L}^2 - \frac{\mathbf{L}^2}{\mu r_0^3}x + \cdots$$

This leads to a perturbation in the rotation-vibration hamiltonian in Eqs. (3.30)

$$\begin{aligned}
H &= H_0 + H' \\
H_0 &= H_{\rm rot} + H_{\rm vib} \\
H' &= -\frac{\mathbf{L}^2}{\mu r_0^3}x
\end{aligned}$$

The eigenfunctions and eigenvalues for this part of the hamiltonian are

$$\begin{aligned}
H_0\psi_{nlm} &= \varepsilon_{nlm}\,\psi_{nlm} \\
\psi_{nlm} &= \psi_n(x)\,\psi_{lm}(\theta,\phi) \\
\varepsilon_{nlm} &= \hbar\omega\left(n+\frac{1}{2}\right) + \frac{\hbar^2}{2I}l(l+1) \qquad ; \ (n,l) = 0,1,2,\cdots \\
& \qquad\qquad\qquad\qquad\qquad\qquad\qquad -l \le m \le l
\end{aligned}$$

The perturbation is diagonal in the degenerate subspace of given (n,l), and

[6] Equivalently, there is no vector $\mathbf{r}$ left in the problem!

first-order perturbation theory then gives (see [Walecka (2013)])

$$\psi_{nlm}^{(1)} = \psi_{nlm} + \sum_{n' \neq n} \frac{\langle n'lm|H'|nlm\rangle}{\hbar\omega(n-n')}\psi_{n'lm}$$
$$\varepsilon_{nlm}^{(1)} = \varepsilon_{nlm} + \langle nlm|H'|nlm\rangle$$

Since x only takes the oscillator up or down one level, this is (see Prob. 3.9)

$$\psi_{nlm}^{(1)} = \psi_{nlm}(x,\theta,\phi) - \frac{\hbar^2 l(l+1)}{\mu r_0^2}\left(\frac{\hbar}{2\mu\omega r_0^2}\right)^{1/2}\frac{1}{\hbar\omega}$$
$$\times\left[-\sqrt{n+1}\,\psi_{n+1}(x) + \sqrt{n}\,\psi_{n-1}(x)\right]\psi_{lm}(\theta,\phi)$$
$$\varepsilon_{nlm}^{(1)} = \varepsilon_{nlm}$$

To this order, the perturbation mixes in the neighboring vibrational states, while leaving the energy eigenvalues unaffected.

Problem 3.12 One way of arriving at the Born-Oppenheimer approximation for the diatomic molecule in Eq. (3.14) is to substitute a wave function ansatz of the form in Eq. (3.46) into the separated internal stationary-state Schrödinger equation, use Eq. (3.13), and then take electronic matrix elements with the wave function $\psi_{\rm el}(\mathbf{r}_i;\mathbf{r})$. Discuss the approximations made in this approach, and indicate how one would go about estimating the size of the correction terms.

Solution to Problem 3.12

The hamiltonian for the diatomic molecule in Fig. 3.1 in the text is given in Eq. (3.12)

$$H = \frac{1}{2}M\dot{\mathbf{R}}^2 + \frac{1}{2}\mu\dot{\mathbf{r}}^2 + H_{\rm el}(\dot{\mathbf{r}}_i,\mathbf{r}_i;\mathbf{r})$$

where $H_{\rm el}(\dot{\mathbf{r}}_i,\mathbf{r}_i;\mathbf{r})$ is written out in detail there. The C-M motion clearly separates

$$\Psi = \psi_{\rm CM}(\mathbf{R})\,\Psi_{\rm tot}(\mathbf{r}_i;\mathbf{r})$$
$$E = \varepsilon^{\rm CM} + \varepsilon^{\rm int}$$

and the time-independent Schrödinger-equation then takes the form

$$\left[\frac{1}{2}\mu\dot{\mathbf{r}}^2 + H_{\rm el}(\dot{\mathbf{r}}_i,\mathbf{r}_i;\mathbf{r})\right]\Psi_{\rm tot}(\mathbf{r}_i;\mathbf{r}) = \varepsilon^{\rm int}\,\Psi_{\rm tot}(\mathbf{r}_i;\mathbf{r})$$

Now assume the wave function has the separated form in Eq. (3.46)

$$\Psi_{\rm tot}(\mathbf{r}_i;\mathbf{r}) = \psi_{\rm el}(\mathbf{r}_i;\mathbf{r})\,\psi_{\rm int}(\mathbf{r})$$

where, from Eq. (3.13),

$$H_{\rm el}(\dot{\mathbf{r}}_i,\mathbf{r}_i;\mathbf{r})\,\psi_{\rm el}(\mathbf{r}_i;\mathbf{r}) = U(r)\,\psi_{\rm el}(\mathbf{r}_i;\mathbf{r})$$

Substitute the assumed form of the wave function into the Schrödinger equation, use the above relation, and then take the matrix element with $\psi_{\rm el}(\mathbf{r}_i;\mathbf{r})$, using the normalization

$$\int d^3r_i\cdots d^3r_n\,|\psi_{\rm el}(\mathbf{r}_i;\mathbf{r})\,|^2 = 1$$

It follows that

$$\int d^3r_i\cdots d^3r_n\,\psi^\star_{\rm el}(\mathbf{r}_i;\mathbf{r})\frac{1}{2}\mu\dot{\mathbf{r}}^2\psi_{\rm el}(\mathbf{r}_i;\mathbf{r})\,\psi_{\rm int}(\mathbf{r}) + U(r)\,\psi_{\rm int}(\mathbf{r}) = \varepsilon^{\rm int}\,\psi_{\rm int}(\mathbf{r})$$

If the $\mu\dot{\mathbf{r}}^2$ can be moved to the left through the electronic wave function, this reduces to the Born-Oppenheimer approximation in Eq. (3.14)

$$\left[\frac{1}{2}\mu\dot{\mathbf{r}}^2 + U(r)\right]\psi_{\rm int}(\mathbf{r}) = H_{\rm int}\psi_{\rm int}(\mathbf{r}) = \varepsilon^{\rm int}\psi_{\rm int}(\mathbf{r})$$

In this case, the assumed separated wave function provides a solution to the time-independent Schrödinger equation.

In fact, the operator $\mu\dot{\mathbf{r}}^2$ does *not* commute with the electronic wave function, and hence there is a correction to the Born-Oppenheimer approximation of[7]

$$\delta H_{\rm int} = \int d^3r_i\cdots d^3r_n\,\left[\psi^\star_{\rm el}(\mathbf{r}_i;\mathbf{r}),\,\frac{1}{2\mu}\mathbf{p}_r^2\right]\psi_{\rm el}(\mathbf{r}_i;\mathbf{r})$$

This expression involves a change in the relative nuclear coordinate $\mathbf{r}$ for a given set of electron coordinates $\mathbf{r}_i$. One can insert an approximate electronic wave function and estimate the magnitude of the correction.[8] The Born-Oppenheimer approximation will be valid if

$$\frac{1}{\varepsilon^{\rm int}}\delta H_{\rm int} \ll 1 \qquad\qquad ;\ \text{validity}$$

Problem 3.13 The dependence on γ, the angle of rotation about the figure axis, of the wave function for a symmetric top in Eq. (3.78) is $e^{i\kappa\gamma}$

[7]Recall $\mu\dot{\mathbf{r}} = \mathbf{p}_r = -i\hbar\boldsymbol{\nabla}_r$.

[8]We do not attempt that here.

(see [Edmonds (1974)]). In general, the quantum number κ describing the angular momentum along the figure axis is restricted by the internal symmetry of the molecule.

(a) Suppose that the molecule is unchanged under rotations of $\gamma \to \gamma + 2\pi/\sigma$ about the figure axis, where σ is a positive integer, and the wave function is required to be *periodic* under such rotations. Show this implies $\kappa = p\sigma$ where $p = 0, \pm 1, \pm 2, \cdots$;

(b) Show the modification of the partition function in Eq. (3.88) is then

$$(\text{p.f.})_{\text{top}} = \frac{\sqrt{\pi}}{\sigma}\left(\frac{8\pi^2 k_{\text{B}}T}{h^2}\right)^{3/2}(I^2 I_3)^{1/2} \qquad ; \ T \to \infty$$

Interpret this result.

Solution to Problem 3.13

(a) The dependence on γ, the angle of rotation about the figure axis, of the wave function for a symmetric top in Eq. (3.78) is $e^{i\kappa\gamma}$.[9] Suppose that the molecule is unchanged under rotations of $\gamma \to \gamma + 2\pi/\sigma$ about the figure axis, where σ is a positive integer, and the wave function is required to be *periodic* under such rotations. This implies

$$\begin{aligned} e^{i\kappa(\gamma+2\pi/\sigma)} &= e^{i\kappa\gamma} \\ e^{2\pi i\kappa/\sigma} &= 1 \\ \kappa/\sigma &= p \qquad\qquad ; \ p = 0, \pm 1, \pm 2, \cdots \end{aligned}$$

(b) The only change in the calculation of the partition function for the symmetric top in Eqs. (3.81)–(3.85) is that the sum over κ now becomes a sum over p, where $\kappa = p\sigma$. Therefore, when the sum over κ is converted to an integral in Eq. (3.86), one has instead

$$\begin{aligned} \mathcal{S} &= 2\int_0^\infty dp \, \exp\left\{-\frac{\hbar^2}{2I_3 k_{\text{B}}T}\left(\kappa + \frac{I_3}{2I}\right)^2\right\} \\ &= \frac{2}{\sigma}\int_0^\infty d\kappa \, \exp\left\{-\frac{\hbar^2}{2I_3 k_{\text{B}}T}\left(\kappa + \frac{I_3}{2I}\right)^2\right\} \end{aligned}$$

Thus Eq. (3.87) becomes

$$\mathcal{S} = \frac{\sqrt{\pi}}{\sigma}\left(\frac{2I_3 k_{\text{B}}T}{\hbar^2}\right)^{1/2} \qquad ; \ T \to \infty$$

[9] See Fig. 3.13 in the text, where the 3-axis is the figure axis.

Hence, Eq. (3.88) now gives for the partition function of this symmetric top

$$(\text{p.f.})_{\text{top}} = \frac{\sqrt{\pi}}{\sigma}\left(\frac{8\pi^2 k_{\text{B}}T}{h^2}\right)^{3/2}\left(I^2 I_3\right)^{1/2} \qquad ; \; T \to \infty$$

The molecule is unchanged under rotations of $2\pi/\sigma$ about the figure axis, and σ is then precisely the *symmetry factor* of section 3.2.2.3 and Eq. (3.93).

Problem 3.14 (a) Show that the enthalpy of a perfect gas is $\mathcal{H} = (5/2)nRT$ (recall Prob. 1.8);

(b) Show that the molar constant-pressure heat capacity of a perfect gas is given by $\mathcal{C}_P = (5/2)R = \mathcal{C}_V + R$.

Solution to Problem 3.14

(a) For a perfect gas, one has from Eqs. (2.133) and (2.136)

$$E = \frac{3}{2}Nk_{\text{B}}T \qquad ; \text{ perfect gas}$$
$$PV = Nk_{\text{B}}T$$

The *enthalpy* is defined in Prob. 1.8

$$\mathcal{H} = E + PV = \frac{5}{2}Nk_{\text{B}}T \qquad ; \text{ enthalpy}$$

Hence the enthalpy of a perfect gas is larger than its energy by $Nk_{\text{B}}T$.

(b) It is shown in Prob. 1.8(a) that in terms of the enthalpy, the first law becomes

$$d\mathcal{H} = đQ + VdP \qquad ; \text{ first law}$$

Therefore, the constant-pressure heat capacity is written in terms of the enthalpy as[10]

$$C_P = \left(\frac{đQ}{dT}\right)_P = \left(\frac{d\mathcal{H}}{dT}\right)_P \qquad ; \text{ given N}$$
$$= \frac{5}{2}Nk_{\text{B}}$$

Hence, for the molar values

$$\mathcal{C}_P = \frac{5}{2}R = \mathcal{C}_V + R$$

[10]Compare Eqs. (2.50).

The molar constant-pressure specific heat of a perfect gas is correspondingly larger than the constant-volume value by R.

Problem 3.15 (a) Directly obtain the result for the induced magnetic moment for spin-1/2 in the first of Eqs. (3.146) and Eq. (3.149) by explicitly carrying out the sums in Eqs. (3.136) and (3.142);

(b) The experimental value of the magnetic moment μ in Eq. (3.135) is defined with respect to J, and it can have either sign. Show that $\langle\mu_F\rangle$ in Eq. (3.146) is even in μ, and hence one can employ $|\mu|$ in discussing it.

Solution to Problem 3.15

(a) For spin-1/2, the expectation value of the magnetic moment in the field F is given in the first of Eqs. (3.142)

$$\langle\mu_F\rangle = \frac{1}{(\text{p.f.})_F}\sum_{m=-1/2}^{1/2}(g\mu_0 m)\exp\left\{\frac{g\mu_0 F m}{k_B T}\right\}$$

If we explicitly carry out the sum over $m = \pm 1/2$, we have

$$\langle\mu_F\rangle = \mu\,\frac{e^{\mu F/k_B T} - e^{-\mu F/k_B T}}{e^{\mu F/k_B T} + e^{-\mu F/k_B T}} \qquad ;\ \mu = \frac{g}{2}\mu_0$$

Here $\mu = g\mu_0/2$ is the experimental moment.[11] This expression is

$$\langle\mu_F\rangle = \mu\tanh y \qquad ;\ y = \frac{\mu F}{k_B T}$$

which is just the first of Eqs. (3.146) and Eq. (3.149).

(b) Since $\coth y = \cosh y/\sinh y$ is odd under $y \to -y$, Eqs. (3.146) give

$$\langle\mu_F\rangle = \mu L_J\left(\frac{\mu F}{k_B T}\right) = -\mu L_J\left(\frac{-\mu F}{k_B T}\right)$$

Thus $\langle\mu_F\rangle$ in Eq. (3.146) is even in μ, and hence one can employ $|\mu|$ in discussing it.

Problem 3.16 Include a contribution of $-\alpha F^2/2$ in the hamiltonian in Eq. (3.117), assume a uniform polarizability α, with no angle or momentum dependence, and show that the analysis of the induced moment arising from the orientation of the dipoles in Eqs. (3.123)–(3.131) is unaffected.

[11] Remember that in section 3.3 μ is the magnetic moment, and *not* the chemical potential.

Solution to Problem 3.16

If a contibution of $-\alpha F^2/2$ is included in the hamiltonian in Eq. (3.117), one has

$$H = \frac{1}{2I}\left(p_\theta^2 + \frac{1}{\sin^2\theta}p_\phi^2\right) - \mu F\cos\theta - \frac{1}{2}\alpha F^2$$

If α has no momentum dependence, then Eq. (3.119) holds with Eq. (3.120) modified to read

$$(\text{p.f.})_F = \frac{1}{2}\int_0^\pi e^{(\mu F\cos\theta + \alpha F^2/2)/k_\text{B}T}\sin\theta\, d\theta$$

Equation (3.123), in turn, becomes

$$\frac{n^\star(\theta)\, d\theta}{N} = \frac{1}{(\text{p.f.})_F}\frac{1}{2}e^{(\mu F\cos\theta + \alpha F^2/2)/k_\text{B}T}\sin\theta\, d\theta$$

If α is independent of angle, the term $e^{\alpha F^2/2k_\text{B}T}$ factors out of the integral in $(\text{p.f.})_F$ and *cancels* in this ratio. The subsequent analysis then proceeds exactly as in the text.

Problem 3.17 Consider two charges $\pm q$ connected by a spring and originally unseparated. Now apply an electric field $\mathbf{E}$, which will stretch the spring and create a dipole $\mathbf{d}$. The polarizability α of this system is defined by $\mathbf{d} = \alpha\mathbf{E}$ where α is a constant (Fig. 3.3).

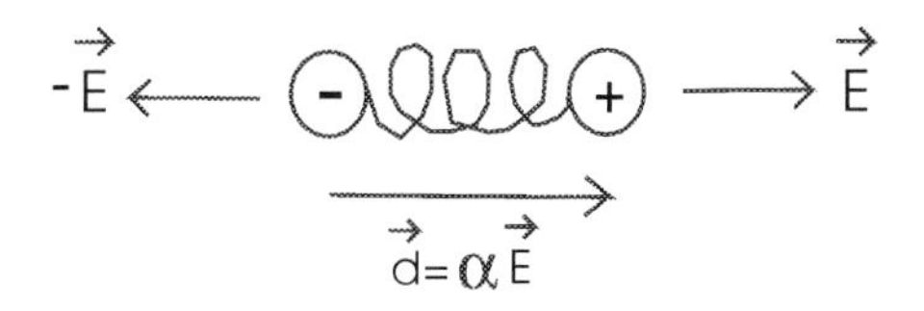

Fig. 3.3 Induced dipole in external electric field modeled with two charges on a spring.

(a) Show that the work done as the electric field is increased by an amount dE is

$$dW = Fds = (E + dE)\alpha[(E + dE) - E] \approx \alpha E dE$$

(b) Integrate this result to show the work required to create the dipole $\mathbf{d}$ is $W = \alpha\mathbf{E}^2/2$;

(c) Show the energy of the induced dipole in the external field is therefore

$$E_{\rm in} = -\mathbf{d}_{\rm ind} \cdot \mathbf{E} + \frac{1}{2}\alpha\mathbf{E}^2 = -\frac{1}{2}\alpha\mathbf{E}^2$$

(d) Show the polarizability of the spring is $\alpha = q^2/k$ where k is the spring constant.

Solution to Problem 3.17

(a) The induced dipole moment $\mathbf{d}_{\rm ind} = \alpha\mathbf{E}$ of the system in Fig. 3.3, relative to the origin, is $\mathbf{d}_{\rm ind} = q\mathbf{s}/2 - q(-\mathbf{s})/2 = q\mathbf{s}$ where $\mathbf{s}$ is the separation of the charges at that field strength

$$\begin{aligned}\mathbf{d}_{\rm ind} &= \alpha\mathbf{E} \\ &= q\frac{\mathbf{s}}{2} - q\frac{(-\mathbf{s})}{2} = q\mathbf{s}\end{aligned}$$

If the field strength is increased to $\mathbf{E} + d\mathbf{E}$, it does work on the charges of

$$dW = q\mathbf{E} \cdot \frac{d\mathbf{s}}{2} - q\mathbf{E} \cdot \frac{(-d\mathbf{s})}{2} = q\mathbf{E} \cdot d\mathbf{s}$$

where we retain only first order in small quantities. It follows from $\alpha\mathbf{E} = q\mathbf{s}$ that the increase in separation is related back to the field by

$$q\, d\mathbf{s} = \alpha\, d\mathbf{E}$$

The differential work done on the charges is therefore

$$dW = q\mathbf{E} \cdot d\mathbf{s} = \alpha\mathbf{E} \cdot d\mathbf{E}$$

(b) The integration of the result in part (a), for constant α, gives

$$W = \frac{1}{2}\alpha\mathbf{E}^2$$

(c) The energy of the induced dipole in the field now consists of two parts—the energy of orientation of the dipole in the field plus the energy stored in its creation

$$E_{\rm in} = -\mathbf{d}_{\rm ind} \cdot \mathbf{E} + \frac{1}{2}\alpha\mathbf{E}^2 = -\frac{1}{2}\alpha\mathbf{E}^2$$

(d) The net force on the positive charge in part (a) vanishes. Thus along $\mathbf{E}$ one has $F_q = qE - ks = 0$, where k is the spring constant and the charges are originally unseparated. Therefore, with a spring, the extension is

$$s = \frac{qE}{k}$$

The induced dipole moment is then

$$d = qs = \frac{q^2E}{k}$$

Hence, the polarizability of the spring is

$$\alpha = \frac{q^2}{k} \qquad ; \text{ spring}$$

Problem 3.18 Consider a paramagnetic sample placed in a solenoid, which produces a uniform magnetic field **H**. A uniform magnetization per unit volume **M** is induced in the sample (Fig. 3.4).

(a) The magnetic field in the material is

$$\mathbf{B} = \mathbf{H} + 4\pi\mathbf{M}$$

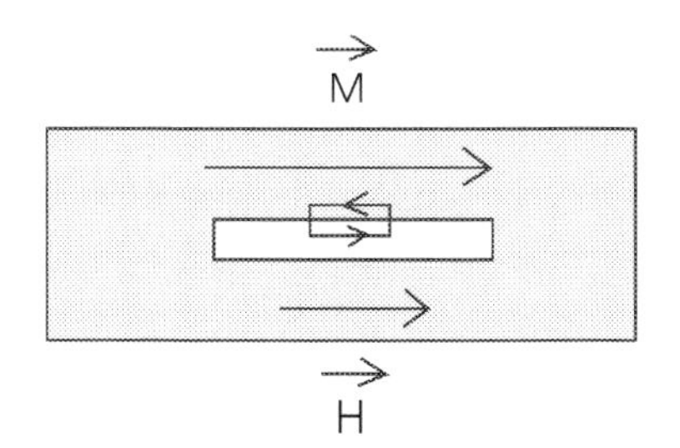

Fig. 3.4 Paramagnetic sample in a solenoid, which produces a uniform magnetic field **H**. A uniform magnetic moment per unit volume **M** is induced in the sample. Also shown is a needle-shaped cavity and a small element of area used to derive a boundary condition.

There is no free current in the material so $\boldsymbol{\nabla} \times \mathbf{H} = 0$ everywhere. Integrate this relation over the little element of area in Fig. 3.4 to show that the tangential component of **H** is unchanged as one moves from the medium into the indicated cavity;

(b) The magnetic field satisfies $\boldsymbol{\nabla} \cdot \mathbf{B} = 0$ everywhere. Take the divergence of the expression in (a), and define a magnetic charge by

$$\rho_m \equiv -\boldsymbol{\nabla} \cdot \mathbf{M} \qquad ; \text{ magnetic charge}$$

Hence show

$$\boldsymbol{\nabla}\cdot\mathbf{H} = 4\pi\rho_m \qquad ; \ \boldsymbol{\nabla}\times\mathbf{H} = 0$$

Use the above results to establish a strict analogy between ($\mathbf{H} \rightleftharpoons \mathbf{E}$), ($\mathbf{M} \rightleftharpoons \mathbf{P}$), and ($\mathbf{B} \rightleftharpoons \mathbf{D}$). Where is ρ_m non-zero in Fig. 3.4?

(c) Now repeat the argument on the polarization in a dielectric medium to obtain the effective magnetic field at the center of a spherical cavity in the paramagnetic material

$$\mathbf{H}_{\text{eff}} = \mathbf{H} + \frac{4\pi}{3}\mathbf{M}$$

(d) Show $\mathbf{B}_{\text{eff}} = \mathbf{H}_{\text{eff}}$ in the cavity.

Solution to Problem 3.18

(a) Consider the small enclosed rectangular area in Fig. 3.4 with surface element $d\mathbf{A}$. There is no free current flowing through it, so

$$\int_A (\boldsymbol{\nabla}\times\mathbf{H})\cdot d\mathbf{A} = 0$$

Now use Stokes' theorem to convert this to a line integral around the boundary.

$$\oint \mathbf{H}\cdot d\mathbf{l} = 0$$

There is only a contribution from the long sides of length dl where $\mathbf{H}$ is antiparallel (parallel) to the boundary and hence

$$\oint \mathbf{H}\cdot d\mathbf{l} = -H_{\text{mat}}\,dl + H_{\text{cav}}\,dl = 0$$

Therefore

$$H_{\text{mat}} = H_{\text{cav}}$$

and the tangential component of $\mathbf{H}$ is unchanged as one moves from the medium into the indicated cavity.

(b) In the material, we have from part (a)[12]

$$\mathbf{B} = \mathbf{H} + 4\pi\mathbf{M}$$

[12]Recall that in this text, we are using c.g.s. units.

Since there is no true magnetic charge, we have from Maxwell's equations

$$\boldsymbol{\nabla}\cdot\mathbf{B}=0$$

It follows that

$$\boldsymbol{\nabla}\cdot\mathbf{H}=-4\pi\,\boldsymbol{\nabla}\cdot\mathbf{M}$$

We can use this relation to define an *effective* magnetic charge ρ_m

$$\rho_m\equiv-\boldsymbol{\nabla}\cdot\mathbf{M}$$

Since $\mathbf{M}$ only exits in the material, we see from Fig. 3.4 that ρ_m will be non-zero at the two ends of the long needle-shaped cavity, where it will have opposite signs.

We have shown that for the configuration in Fig. 3.4

$$\boldsymbol{\nabla}\cdot\mathbf{H}=4\pi\rho_m \qquad ; \ \boldsymbol{\nabla}\times\mathbf{H}=0$$

These equations bear a direct relation to electrostatics where

$$\boldsymbol{\nabla}\cdot\mathbf{E}=4\pi\rho_e \qquad ; \ \boldsymbol{\nabla}\times\mathbf{E}=0$$

Thus we have the analogy ($\mathbf{E}\rightleftharpoons\mathbf{H}$).

(c) This analogy is extended in footnote 40 on p. 110, which observes the following: For a *dielectric* material in a capacitor, and a *paramagnetic* material in a solenoid,

$$\mathbf{D}=\mathbf{E}+4\pi\mathbf{P} \qquad ; \ \mathbf{B}=\mathbf{H}+4\pi\mathbf{M} \qquad ; \ \text{in medium}$$

Here $\mathbf{D}$ is determined from the free charge on the plates, $\mathbf{H}$ is determined from the free current in the wire, and $(\mathbf{E},\mathbf{B})$ are the true fields that produce the Lorentz force. The polarization $\mathbf{P}$ enhances the electric field $\mathbf{D}$, while the magnetization $\mathbf{M}$ similarly enhances the magnetic field $\mathbf{B}$. Thus one has the strict analogy ($\mathbf{D} \rightleftharpoons \mathbf{B}$), ($\mathbf{E} \rightleftharpoons \mathbf{H}$), and ($\mathbf{P} \rightleftharpoons \mathbf{M}$) [compare Eqs. (3.151) and (3.170)].[13]

We can thus repeat the derivation of the Clausius-Mosotti relation in the text using this correspondence, and (effective) magnetic charge rather than electric charge. This leads to the following analogous relations for spherical *cavities* in these materials

$$\mathbf{E}_{\text{eff}} = \mathbf{E} + \frac{4\pi}{3}\mathbf{P} \qquad ; \ \mathbf{H}_{\text{eff}} = \mathbf{H} + \frac{4\pi}{3}\mathbf{M} \qquad ; \text{ in cavity}$$

(d) Since the magnetization $\mathbf{M}$ vanishes in the cavity, we have from the above that

$$\mathbf{B}_{\text{eff}} = \mathbf{H}_{\text{eff}} \qquad ; \text{ in cavity}$$

Problem 3.19 Consider the following chemical reaction between perfect gases

$$A_2 + B_2 \rightleftharpoons 2AB$$

(a) Show the law of mass action and equilibrium constant in this case are given by

$$\frac{N_{AA}N_{BB}}{[N_{AB}]^2} = \frac{(\text{p.f.})_{\text{AA}}(\text{p.f.})_{\text{BB}}}{[(\text{p.f.})_{\text{AB}}]^2} \equiv K_{\text{eq}}(T)$$

where $K_{\text{eq}}(T)$ is independent of V.

(b) Show the chemical potentials satisfy

$$\mu_1 + \mu_2 = 2\mu_{12}$$

[13] Make a gaussian pillbox on the surface of the left end of the cavity in Fig. 3.4, and use Gauss's theorem

$$\int_V (\boldsymbol{\nabla} \cdot \mathbf{M})\, dV = \int_A \mathbf{M} \cdot d\mathbf{A}$$

$$\Longrightarrow \qquad -\rho_m t A \equiv -\sigma_m A = -MA$$

We then have the further analogy [see Eq. (3.155)]

$$\mathbf{M} = \sigma_m\, \mathbf{n} \qquad ; \ \mathbf{P} = \sigma_e\, \mathbf{n} \qquad ; \text{ on surface}$$

Solution to Problem 3.19

We show how the analysis in section 3.4.2 is modified in the case of the following chemical reaction between perfect gases

$$A_2 + B_2 \rightleftharpoons 2AB$$

(a) The steepest descent analysis proceeds as before, with three independent species $(N_1, N_2, N_{12}) = (N_{AA}, N_{BB}, N_{AB})$. The constraint Eqs. (3.196) are then modified to read

$$\begin{aligned} 2N_1 + N_{12} &= N_A \\ 2N_2 + N_{12} &= N_B \end{aligned}$$

where (N_A, N_B) are the number of atoms of type A and B respectively. The new variations in Eqs. (3.199) therefore satisfy

$$\begin{aligned} 2\,\delta N_1 + \delta N_{12} &= 0 \\ 2\,\delta N_2 + \delta N_{12} &= 0 \end{aligned}$$

Identification of the largest term in the sum for Ω, and the use of Lagrange's method of undetermined multipliers, then lead to the modified form of Eqs. (3.200)

$$\begin{aligned} \ln(\text{p.f.})_{\text{AA}} - \ln N_1 + 2\alpha_A &= 0 \\ \ln(\text{p.f.})_{\text{BB}} - \ln N_2 + 2\alpha_B &= 0 \\ \ln(\text{p.f.})_{\text{AB}} - \ln N_{12} + (\alpha_A + \alpha_B) &= 0 \end{aligned}$$

Hence Eqs. (3.202) become

$$\begin{aligned} N_1 &= \lambda_A^2(\text{p.f.})_{\text{AA}} \\ N_2 &= \lambda_B^2(\text{p.f.})_{\text{BB}} \\ N_{12} &= \lambda_A\lambda_B(\text{p.f.})_{\text{AB}} \end{aligned}$$

where $\lambda_A = e^{\alpha_A}$ and $\lambda_B = e^{\alpha_B}$. The new equilibrium constant now follows directly[14]

$$\frac{N_{AA}N_{BB}}{[N_{AB}]^2} = \frac{(\text{p.f.})_{\text{AA}}(\text{p.f.})_{\text{BB}}}{[(\text{p.f.})_{\text{AB}}]^2} \equiv K_{\text{eq}}(T)$$

There is one particle number, and one (p.f.), for each species in the reaction

$$A_2 + B_2 \rightleftharpoons AB + AB$$

[14] See Prob. 3.21 for a discussion of the V-dependence of this expression.

(b) The expression for the Helmholtz free energy in Eq. (3.209) remains unchanged, as does the calculation of the chemical potentials in Eqs. (3.212). All that changes is the last set of equalties in the first two relations in Eqs. (3.212), which follow from the above,

$$\mu_1 = \left(\frac{\partial A}{\partial N_1}\right)_{T,V,N_2,N_{12}} = -k_BT[\ln(\text{p.f.})_{AA} - \ln N_1] = 2\alpha_A k_BT$$

$$\mu_2 = \left(\frac{\partial A}{\partial N_2}\right)_{T,V,N_1,N_{12}} = -k_BT[\ln(\text{p.f.})_{BB} - \ln N_2] = 2\alpha_B k_BT$$

$$\mu_{12} = \left(\frac{\partial A}{\partial N_{12}}\right)_{T,V,N_1,N_2} = -k_BT[\ln(\text{p.f.})_{AB} - \ln N_{12}] = (\alpha_A + \alpha_B)k_BT$$

Hence the chemical potentials now satisfy

$$\mu_1 + \mu_2 = 2\mu_{12}$$

There is one chemical potential for each species in the reaction

$$A_2 + B_2 \rightleftharpoons AB + AB$$

Problem 3.20 (a) Use the result in Prob. 3.19, and find an expression for the equilibrium constant $K_{\text{eq}}(T)$ for the following reaction[15]

$$H_2 + I_2 \rightleftharpoons 2HI$$

in terms of the partition functions for these diatomic molecules.

(b) Compute $K_{\text{eq}}(T)$ at 500°K from the following table of constants

Table 3.1 Molecular constants.

	H_2	I_2	HI
D	36,436	12,625	24,944
ω	4,395	214.6	2,309
r	0.7417	2.667	1.604

The quantities appearing in this table are

$$D = \text{dissociation energy in cm}^{-1}$$
$$\omega = \text{oscillator angular frequency in cm}^{-1}$$
$$r = \text{internuclear spacing in Å}$$

[15]This problem takes a little more work, but it may be the most instructive of all.

(c) Compare with the experimental value of the equilibrium constant at that temperature[16]

$$\log_{10}\left[K_{\rm eq}\right]^{-1} = 2.078 \qquad ; \text{ experiment}$$

Solution to Problem 3.20

(a) The equilibrium constant for the reaction

$$\mathrm{H_2 + I_2 \rightleftharpoons 2HI}$$

is given in terms of the partition functions for these diatomic molecules in Prob. 3.19

$$\frac{N_{AA}N_{BB}}{[N_{AB}]^2} = \frac{(\text{p.f.})_{\rm AA}(\text{p.f.})_{\rm BB}}{[(\text{p.f.})_{\rm AB}]^2} \equiv K_{\rm eq}(T)$$

The partition function for the generic diatomic molecule labeled (12) is given in Eqs. (3.31) and (3.1)–(3.3)

$$(\text{p.f.}) = (\text{p.f.})_{\rm trans}(\text{p.f.})_{\rm rot}(\text{p.f.})_{\rm vib}(\text{p.f.})_{\rm el}$$

The translation partition function follows from Eq. (3.32)

$$(\text{p.f.})_{\rm trans} = V\left[\frac{2\pi(m_1+m_2)k_{\rm B}T}{h^2}\right]^{3/2}$$

If $T/\theta_{\rm R} \gg 1$, which we assume to be the case,[17] then the rotation partition is given in Eqs. (3.39) and (3.45)

$$(\text{p.f.})_{\rm rot} = \frac{8\pi^2 I_{12}\, k_{\rm B}T}{\sigma_{12}\, h^2} \qquad ; \; \frac{T}{\theta_{\rm R}} \gg 1$$

Here σ is the symmetry factor for the molecule [see Eqs. (3.45) and (3.55)], and I is the moment of inertia in Eq. (3.26), with μ the reduced mass

$$I_{12} = \left(\mu r_0^2\right)_{12} = \frac{m_1 m_2}{m_1+m_2}(r_0^2)_{12}$$

The vibration partition function is given in Eqs. (3.34)

$$(\text{p.f.})_{\rm vib} = \frac{e^{-h\nu/2k_{\rm B}T}}{1-e^{-h\nu/k_{\rm B}T}}$$

[16]There is an unfortunate misprint in the text, corrected here; the initial entry in the table should be 36,436 (thanks to *Chris Tennant*, an excellent student at W&M).

[17]Figures 3.7–3.8 in the text imply this should be good approximation for the participants in the given reaction at $T = 500\,^{\rm o}$K, where it will be applied in part (b). Compare Eq. (3.60).

The electronic partition function for the electrons in their ground state is given in Eq. (3.41)

$$(\text{p.f.})_{\text{el}} = e^{-\varepsilon_0/k_{\text{B}}T}$$

Here $\varepsilon_0 \equiv -\omega$ is the energy of the bound electronic ground state of the molecule relative to the zero of energy, which is that of the separated neutral atoms in their ground states [see Fig. 3.28 in the text and Eq. (3.211)].

Substitution into the expression for the equilibrium constant of the given reaction, and cancellation of common factors, gives

$$\begin{aligned} K_{\text{eq}}(T) = & \, 2\frac{(m_A m_B)^{1/2}}{m_A + m_B} \frac{(r_0^2)_{AA}\,(r_0^2)_{BB}}{[(r_0^2)_{AB}]^2} \frac{\sigma_{AB}^2}{\sigma_{AA}\,\sigma_{BB}} \\ & \times \frac{e^{-h\nu_{AA}/2k_{\text{B}}T}}{1 - e^{-h\nu_{AA}/k_{\text{B}}T}} \frac{e^{-h\nu_{BB}/2k_{\text{B}}T}}{1 - e^{-h\nu_{BB}/k_{\text{B}}T}} \left[\frac{e^{-h\nu_{AB}/2k_{\text{B}}T}}{1 - e^{-h\nu_{AB}/k_{\text{B}}T}}\right]^{-2} \\ & \times e^{(2\varepsilon_{AB} - \varepsilon_{AA} - \varepsilon_{BB})/k_{\text{B}}T} \end{aligned}$$

(b) To put in numbers using the values given in Table 3.1, we need to make use of the relations in Eqs. (3.59)

$$\frac{\Delta\varepsilon}{hc} = \lambda^{-1} \qquad ; \; 1\,\text{eV} = 8,066\,\text{cm}^{-1}$$
$$k_{\text{B}} = 8.620 \times 10^{-5}\,\text{eV}/^{\text{o}}\text{K}$$

The conversion factor is written alternatively as

$$hc = \frac{1\,\text{eV}}{8066\,\text{cm}^{-1}}$$

Let us start evaluating each of the contributions to the equilibrium constant, in turn:

(1) The dimensionless ratio in the mass term is evaluated from the atomic masses

$$m_{\text{H}} = 1.0079 \qquad ; \; m_{\text{I}} = 126.9045$$

This gives

$$2\frac{(m_A m_B)^{1/2}}{m_A + m_B} = 0.1768$$

(2) The required dimensionless ratio of radial separations follows directly from Table 3.1

$$\frac{(r_0^2)_{AA}\,(r_0^2)_{BB}}{[(r_0^2)_{AB}]^2} = 0.5911$$

(3) The symmetry factors are

$$\sigma_{AA} = \sigma_{BB} = 2 \qquad\qquad ; \ \sigma_{AB} = 1$$

Thus

$$\frac{\sigma_{AB}^2}{\sigma_{AA}\,\sigma_{BB}} = 0.25$$

(4) At $T = 500\,^{\circ}\mathrm{K}$, one has from the above

$$k_{\mathrm{B}}T = 0.0431\,\mathrm{eV}$$

The dissociation energies are given by

$$\begin{aligned}\varepsilon_A + \varepsilon_B - \varepsilon_{AB} &= hcD_{AB} \\ \varepsilon_A + \varepsilon_A - \varepsilon_{AA} &= hcD_{AA} \\ \varepsilon_B + \varepsilon_B - \varepsilon_{BB} &= hcD_{BB}\end{aligned}$$

It follows that

$$2\varepsilon_{AB} - \varepsilon_{AA} - \varepsilon_{BB} = -hc(2D_{AB} - D_{AA} - D_{BB})$$

This gives from Table 3.1

$$2\varepsilon_{AB} - \varepsilon_{AA} - \varepsilon_{BB} = -0.1025\,\mathrm{eV}$$

The electronic contribution is therefore

$$e^{(2\varepsilon_{AB} - \varepsilon_{AA} - \varepsilon_{BB})/k_{\mathrm{B}}T} = 9.266 \times 10^{-2}$$

(5) The oscillator angular frequency is $\omega = 2\pi\nu$, and what is tabulated in Table 3.1 in cm^{-1} is ω/c. Thus the oscillator energy is $h\nu = (hc/2\pi)(\omega/c)$. The quantities $h\nu_{12}$ needed for the vibrational contribution to the equilibrium constant follow as

$$\begin{aligned}h\nu_{AA} &= 8.672 \times 10^{-2}\,\mathrm{eV} \\ h\nu_{BB} &= 4.234 \times 10^{-3}\,\mathrm{eV} \\ h\nu_{AB} &= 4.556 \times 10^{-2}\,\mathrm{eV}\end{aligned}$$

This implies

$$\frac{h\nu_{AA}}{k_{\rm B}T} = 2.012 \qquad ; \frac{h\nu_{BB}}{k_{\rm B}T} = 0.0982 \qquad ; \frac{h\nu_{AB}}{k_{\rm B}T} = 1.057$$

The vibrational contribution is therefore

$$\frac{e^{-h\nu_{AA}/2k_{\rm B}T}}{1-e^{-h\nu_{AA}/k_{\rm B}T}} \frac{e^{-h\nu_{BB}/2k_{\rm B}T}}{1-e^{-h\nu_{BB}/k_{\rm B}T}} \left[\frac{e^{-h\nu_{AB}/2k_{\rm B}T}}{1-e^{-h\nu_{AB}/k_{\rm B}T}}\right]^{-2} = 5.265$$

A combination of these results gives a calculated value of the equilibrium constant for the reaction $H_2 + I_2 \rightleftharpoons 2HI$ at a temperature of $T = 500^{\rm o}{\rm K}$

$$K_{\rm eq} = 1.275 \times 10^{-2}$$
$$\log_{10}[K_{\rm eq}]^{-1} = 1.895 \qquad ; \text{calculated}$$

(b) The calculated value above should be compared with the experimental value[18]

$$\log_{10}[K_{\rm eq}]^{-1} = 2.078 \qquad ; \text{experiment}$$

These agree to better than 10%. Improvements in the calculated value would include:

- A better treatment of $(\text{p.f.})_{\rm rot}$ at finite T;
- Non-harmonic corrections to the potential $U(r)$;
- Inclusion of rotation-vibration coupling;
- Corrections to the Born-Oppenheimer approximation; *etc.*

At least to this author, it is a truly impressive result from statistical mechanics that given a few physical parameters for the molecules involved, one can calculate the equilibrium constant for the chemical reaction.

Problem 3.21 Consider all the participants in a chemical reaction to be perfect gases. Show that the volume dependence in the equilibrium constant $K(T,V)$ is just such as to allow one to re-express the law of mass action in terms of the particle densities $n_i = N_i/V$.[19]

[18]The experimental value is obtained from the *Handbook of Chemistry and Physics*, as is Table 3.1.

[19]These are often referred to as the "concentrations".

Solution to Problem 3.21

Let us see how this works in the solution to Prob. 3.19. The partition function for an ideal gas in Eq. (2.130) is proportional to the volume

$$(\text{p.f.}) \propto V \qquad ; \text{ ideal gas}$$

Hence the equilibrium constant in Prob. 3.19 is independent of the volume, as claimed

$$\frac{N_{AA}N_{BB}}{[N_{AB}]^2} = \frac{(\text{p.f.})_{\text{AA}}(\text{p.f.})_{\text{BB}}}{[(\text{p.f.})_{\text{AB}}]^2} \equiv K_{\text{eq}}(T)$$

Now divide the numerator and denominator of the l.h.s. by V^2, and this becomes a relation on *concentrations*

$$\frac{n_{AA}\, n_{BB}}{[n_{AB}]^2} = K_{\text{eq}}(T) \qquad ; \; n_i \equiv \frac{N_i}{V}$$

From the analysis in Prob. 3.19, there will always be an appropriate number of volume factors in the law of mass action to perform this conversion for ideal gases.[20]

Problem 3.22 (a) Assume the model of a solid used in Einstein's theory of the specific heat, and treat the vapor as a perfect gas. Derive an explicit expression for the vapor pressure $g(T)^{-1}$ in Eq. (3.232);

(b) Sketch the T-dependence of $g(T)^{-1}$. Discuss.

Solution to Problem 3.22

(a) The *vapor pressure* of a pure solid in equilibrium with its gas is given in Eq. (2.232)

$$P = \frac{k_{\text{B}}T}{V}\frac{(\text{p.f.})_{(1)}}{(\text{p.f.})_{(2)}} \equiv g(T)^{-1} \qquad ; \text{ vapor pressure}$$

Here (1) denotes the gas and (2) the solid. The partition function for a perfect gas is given in Eq. (2.130)

$$(\text{p.f.})_{(1)} = V\left(\frac{2\pi m k_{\text{B}}T}{h^2}\right)^{3/2}$$

We employ a two-dimensional simple harmonic oscillator potential to localize the particle on the surface of the solid. The partition function in this

[20] For the reaction analyzed in the text $A + B \rightleftharpoons AB$, the equilibrium constant will behave as $K_{\text{eq}}(V,T) \propto V$, and that factor of V can then be taken over to the l.h.s. to convert it to concentrations.

version of Einstein's model then follows from Eq. (2.46)

$$(\text{p.f.})_{(2)} = \left[\frac{e^{-h\nu_0/2k_\text{B}T}}{1 - e^{-h\nu_0/k_\text{B}T}}\right]^2$$

where ν_0 is the basic oscillator frequency. Hence, the vapor pressure is

$$P = g(T)^{-1} = 4\,k_\text{B}T\left(\frac{2\pi m k_\text{B}T}{h^2}\right)^{3/2}\sinh^2\left(\frac{h\nu_0}{2k_\text{B}T}\right)$$

At high temperature, as long as the solid remains a solid, this is

$$g(T)^{-1} = \frac{(h\nu_0)^2}{k_\text{B}T}\left(\frac{2\pi m k_\text{B}T}{h^2}\right)^{3/2} \qquad ; T \to \infty$$

At low temperature, as long as the gas remains a gas, this is

$$g(T)^{-1} = k_\text{B}T\left(\frac{2\pi m k_\text{B}T}{h^2}\right)^{3/2} e^{h\nu_0/k_\text{B}T} \qquad ; T \to 0$$

(b) The vapor pressure can be re-written in terms of $\theta_E = h\nu_0/k_\text{B}$ [21]

$$P = 4(h\nu_0)^{5/2}\left(\frac{2\pi m}{h^2}\right)^{3/2}\left(\frac{T}{\theta_E}\right)^{5/2}\sinh^2\left(\frac{\theta_E}{2T}\right) \qquad ; \theta_E = \frac{h\nu_0}{k_\text{B}}$$

(b) Define

$$P \equiv 4(h\nu_0)^{5/2}\left(\frac{2\pi m}{h^2}\right)^{3/2} F\left(\frac{T}{\theta_\text{E}}\right)$$
$$F(t) = t^{5/2}\sinh^2(1/2t)$$

Readers can convince themselves that the constants in front have the dimension of force/area, or pressure, and the temperature dependence is then contained in the dimensionless function $F(T/\theta_E)$, which is plotted in Fig. 3.5.

The exponential growth at very small T/θ_E arises from the one-body harmonic-oscillator model of the solid with a single frequency ν_0, which oversimplifies the normal modes.[22] Furthermore, this temperature region is presumably unphysical for the vapor pressure since the gas is unlikely

[21] See Prob. 5.6.
[22] Compare Prob. 7.5.

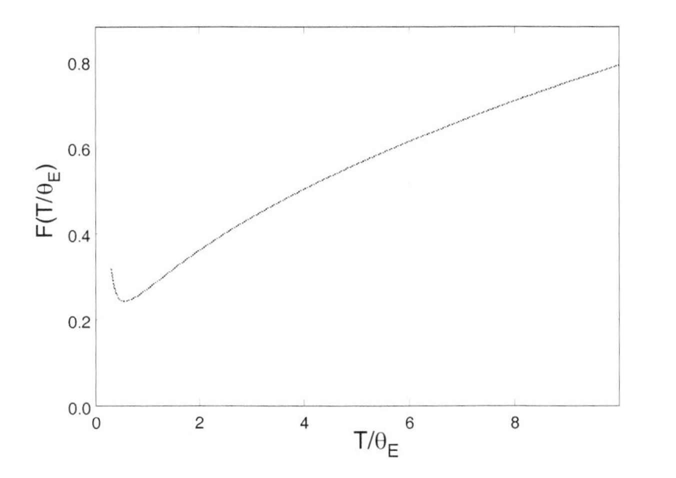

Fig. 3.5 Temperature dependence of the vapor pressure of a solid in equilibrium with its vapor, based on the Einstein model of the solid with $\theta_E = h\nu_0/k_B$. Here the pressure is $P = 4(h\nu_0)^{5/2}\,(2\pi m/h^2)^{3/2}\,F\,(T/\theta_E)$, where the dimensionless function $F(t)$ is $F(t) = t^{5/2}\sinh^2(1/2t)$.

to remain a vapor down there. In addition, the application of classical statistics at very low temperature is inappropriate.

A more sophisticated model of $(\text{p.f.})_{(2)}$ is examined in Prob. 3.24, adding the potential in Fig. 3.6 in a direction perpendicular to the surface of the solid. Readers are urged to use the results in Prob. 3.24 to make some improved plots of the temperature dependence of the vapor pressure of a solid in equilibrium with its vapor.

Problem 3.23 Go through the arguments in the method of steepest descent in detail, and verify Eqs. (3.187) and (3.190).[23]

Solution to Problem 3.23

The goal is to find the coefficient of z^E in Eq. (3.183), for this coefficient provides the appropriate number of complexions $\Omega(N_1, N_2, \cdots; E)$ for an assembly composed of a fixed number of the distinct systems. The systems are non-localized, or localized, according to the factors of $1/N_i!$ retained in Eq. (3.183). For clarity, we temporarily suppress the explicit V dependence

[23] *Hint*: Leave the steepest-descent result in the form of Eq. (2.85).

in Ω. The method of steepest descent solves for this coefficient as

$$\Omega(N_1, N_2, \cdots; E) = \frac{1}{N_1!}\frac{1}{N_2!}\cdots\frac{1}{2\pi i}\oint\frac{dz}{z^{E+1}}[f^{(1)}(z)]^{N_1}[f^{(2)}(z)]^{N_2}\cdots$$

Write $N_i = \xi_i N$ where ξ_i is a finite fraction, and consider the limit $N \to \infty$. In the vicinity of the saddle point, the integrand becomes

$$\frac{\{[f^{(1)}(z)]^{\xi_1}[f^{(2)}(z)]^{\xi_2}\cdots\}^N}{z^{E+1}} \equiv e^{Ng(z)}$$

$$g(z) = \ln\{[f^{(1)}(z)]^{\xi_1}[f^{(2)}(z)]^{\xi_2}\cdots\} - \frac{E}{N}\ln z$$

where we have neglected the term of $O(1/N)$.

Now proceed exactly as in Eqs. (2.79)–(2.85) in the text, to obtain

$$\frac{1}{2\pi i}\oint dz\, e^{Ng(z)} = \frac{e^{Ng(x_0)}}{[2\pi Ng''(x_0)]^{1/2}}$$

Hence, with $\ln N \ll N$,

$$\begin{aligned}\ln\left[\frac{1}{2\pi i}\oint dz\, e^{Ng(z)}\right] &= Ng(x_0)\\ &= \ln\{[f^{(1)}(x_0)]^{\xi_1}[f^{(2)}(x_0)]^{\xi_2}\cdots\}^N - E\ln x_0\end{aligned}$$

This gives

$$\ln\Omega(N_1, N_2, \cdots; E) = \ln\left\{\frac{[f^{(1)}(x_0)]^{N_1}[f^{(2)}(x_0)]^{N_2}\cdots}{N_1!N_2!\cdots}\right\} - E\ln x_0$$

which is Eq. (3.187).

The saddle point $x_0 \equiv e^\beta$ is now located exactly as in Eqs. (3.188)–(3.190), with the result

$$N_1\frac{\sum_i \varepsilon_i^{(1)} e^{\beta\varepsilon_i^{(1)}}}{\sum_i e^{\beta\varepsilon_i^{(1)}}} + N_2\frac{\sum_i \varepsilon_i^{(2)} e^{\beta\varepsilon_i^{(2)}}}{\sum_i e^{\beta\varepsilon_i^{(2)}}} + \cdots = E$$

This relation implicitly determines $\beta(N_1, N_2, \cdots, E)$.[24]

Problem 3.24 (a) Make a model of a system on a surface site as a particle bound in a two-dimensional harmonic oscillator in the transverse directions, and in a square-well potential in the direction perpendicular to the surface. Compute the ratio of partition functions $g(T)$ in Eq. (3.246),

[24] In Eq. (3.208), we later determine $\beta = -1/k_BT$.

and obtain an explicit expression for the Langmuir adsorption isotherm in Eq. (3.248);

(b) Discuss the validity of this model. How would you improve it?

Solution to Problem 3.24

(a) The Langmuir adsorption isotherm is given in Eq. (3.248) and sketched in Fig. 3.30 in the text. The quantity $g(T)$ appearing in the isotherm is defined in Eq. (3.246)

$$\frac{k_{\rm B}T}{V}\frac{({\rm p.f.})_{(1)}}{({\rm p.f.})_{(2)}} \equiv \frac{1}{g(T)}$$

where

- $({\rm p.f.})_{(1)}$ is the partition function for the gas;
- $({\rm p.f.})_{(2)}$ is the partition function for a system localized on a site in the surface.

From Eq. (2.130), the partition function for the gas is

$$({\rm p.f.})_{(1)} = V\left(\frac{2\pi m k_{\rm B}T}{h^2}\right)^{3/2}$$

Hence

$$\left(\frac{2\pi m}{h^2}\right)^{3/2}\frac{(k_{\rm B}T)^{5/2}}{({\rm p.f.})_{(2)}} = \frac{1}{g(T)}$$

It remains to model $({\rm p.f.})_{(2)}$.

Since the energies in the three orthogonal modes are additive as in Eq. (3.1), the partition function for a system on the surface factors as in Eq. (3.3). The two transverse modes in the (y, z)-directions each give the harmonic oscillator result in Eq. (2.46),[25] and therefore

$$({\rm p.f.})_{(2)} = \left(\frac{e^{-h\nu/2k_{\rm B}T}}{1-e^{-h\nu/k_{\rm B}T}}\right)^2 ({\rm p.f.})_{\perp}$$

where $({\rm p.f.})_{\perp}$ now only involves a sum over modes in the x-direction, perpendicular to the surface

$$({\rm p.f.})_{\perp} = \sum_i e^{-\varepsilon_i/k_{\rm B}T} \qquad ; \; x\text{-direction}$$

[25] Recall Prob. 3.22.

Let us model the perpendicular potential as given in the problem, and shown in Fig. 3.6.

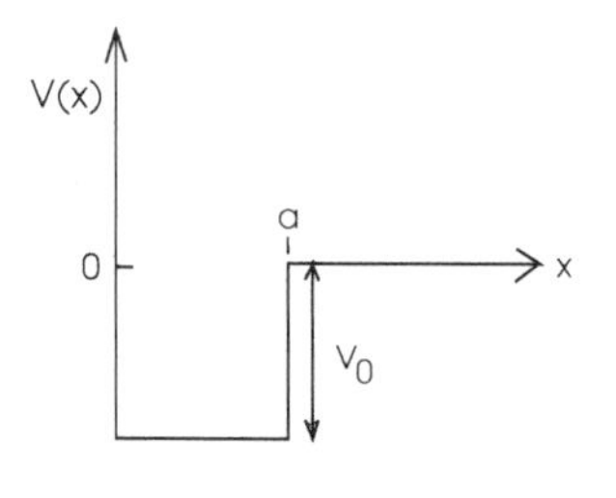

Fig. 3.6 Model potential in the x-direction perpendicular to the surface.

This one-dimensional problem is solvable. If the potential is deep enough, the first few bound states are given by

$$\varepsilon_n = -V_0 + \frac{\hbar^2\pi^2}{2ma^2}n^2 \qquad ; \; n = 1, 2, 3, \cdots$$

Hence, in the limit $k_{\rm B}T \ll V_0$, the transverse partition function is given by

$$(\text{p.f.})_\perp \approx e^{V_0/k_{\rm B}T}\left[e^{-\hbar^2\pi^2/2ma^2k_{\rm B}T} + e^{-4\hbar^2\pi^2/2ma^2k_{\rm B}T} + \cdots\right] \quad ; \; \frac{k_{\rm B}T}{V_0} \ll 1$$

More generally, there is also a continuum contribution from the model potential in Fig. 3.6. If the continuum states at an energy $\varepsilon = \hbar^2k^2/2m$ are approximated by free particles in a box of size $L = V^{1/3}$, then the $(\text{p.f.})_\perp$ is given approximately by

$$(\text{p.f.})_\perp \approx e^{V_0/k_{\rm B}T}\left[e^{-\hbar^2\pi^2/2ma^2k_{\rm B}T} + e^{-4\hbar^2\pi^2/2ma^2k_{\rm B}T} + \cdots\right] + \sum_{n=1}^{\infty} e^{-\hbar^2\pi^2n^2/2mL^2k_{\rm B}T}$$

With this approximation for $(\text{p.f.})_\perp$, the function $g(T)$ is given by

$$g(T) \approx \left(\frac{h^2}{2\pi m}\right)^{3/2}\left(\frac{e^{-h\nu/2k_{\rm B}T}}{1 - e^{-h\nu/k_{\rm B}T}}\right)^2 \frac{(\text{p.f.})_\perp}{(k_{\rm B}T)^{5/2}}$$

If we write $L = V^{1/3}$, then the continuum contribution puts a hidden volume dependence in $g(T)$.

(b) There are various improvements one can think of here:

- Within the microcanonical ensemble:
  - Use the full set of exact one-body solutions in the potential in Fig. 3.6 in computing $(\text{p.f.})_\perp$;
  - Use more realistic potentials in the transverse and perpendicular directions;
- Go to the canonical ensemble:
  - Use a set of lattice gas potentials in the transverse direction (see Prob. 8.5) together with a tethering potential in the perpendicular direction. This treats the surface particles as a many-body problem;
  - Of course, one can then also treat the gas phase as an imperfect gas.

Problem 3.25 Add explicit subscripts to indicate the variables that are to be held fixed in computing each of the partial derivatives in Eqs. (3.215)–(3.217).

Solution to Problem 3.25

Solution to this problem just involves judicious use of the chain rule for differentiation. Equations (3.215), with subscripts added, read

$$
\begin{aligned}
\mu_A &= \left[\frac{\partial A(T,V,N_1,N_2,N_{12})}{\partial N_A}\right]_{T,V,N_B} \\
&= \left(\frac{\partial A}{\partial N_1}\right)_{T,V,N_2,N_{12}}\left(\frac{\partial N_1}{\partial N_A}\right)_{N_B} + \left(\frac{\partial A}{\partial N_2}\right)_{T,V,N_1,N_{12}}\left(\frac{\partial N_2}{\partial N_A}\right)_{N_B} \\
&\quad + \left(\frac{\partial A}{\partial N_{12}}\right)_{T,V,N_1,N_2}\left(\frac{\partial N_{12}}{\partial N_A}\right)_{N_B}
\end{aligned}
$$

Equations (3.216) then read

$$
\begin{aligned}
\left(\frac{\partial N_1}{\partial N_A}\right)_{N_B} + \left(\frac{\partial N_{12}}{\partial N_A}\right)_{N_B} &= 1 \\
\left(\frac{\partial N_2}{\partial N_A}\right)_{N_B} + \left(\frac{\partial N_{12}}{\partial N_A}\right)_{N_B} &= 0
\end{aligned}
$$

With the use of Eqs. (3.212), Eqs. (3.217) then become

$$\mu_A = \mu_1 + (-\mu_1 - \mu_2 + \mu_{12}) \left(\frac{\partial N_{12}}{\partial N_A} \right)_{N_B} = \mu_1$$

where Eq. (3.213) has again been employed.

This establishes the result that the chemical potential for a substance treated as a component is identical to the chemical potential of that substance treated as a species.

Chapter 4

The Canonical Ensemble

Problem 4.1 (a) Show that the error in Eq. (4.40) is indeed of $O(1/n)$;

(b) Show that Eq. (4.41) is then Stirling's formula;

(c) Show that with the neglect of terms of $O(1/n)$, Eq. (4.41) provides an analytic continuation of Stirling's formula to non-integer n.

Solution to Problem 4.1

(a) We start from Eq. (4.36)

$$\int_0^\infty x^{n-1}\, e^{n(1-x)}\, dx \equiv \int_0^\infty I(x)dx$$

For large n, the logarithm of the integrand $I(x)$ takes the form in Eq. (4.37)

$$\ln I(x) = n[\ln x + 1 - x] \leq 0 \qquad ; \; n \to \infty$$

Note this already neglects a term of relative $O(1/n)$ in the exponent (see below).

The integrand $I(x)$ is sharply peaked about $x = 1$ (See Fig. 4.3 in the text). If $\ln I(x)$ is expanded in a Taylor series through $O[(1-x)^2]$, the result is Eqs. (4.40)

$$\begin{aligned}\int_0^\infty I(x)dx &= \int_0^\infty e^{\ln I(x)}\, dx \\ &\approx \int_0^\infty e^{-(x-1)^2 n/2}\, dx \approx \int_{-\infty}^\infty e^{-nu^2/2}\, du\end{aligned}$$

There are two approximations here, and we first examine the second one. Since $u = (x-1)$, the extention of the integral to $-\infty$ adds the following contribution

$$\delta \int_0^\infty I(x)dx = \int_{-\infty}^{-1} e^{-nu^2/2}\, du$$
$$= \frac{1}{\sqrt{n}} \int_{-\infty}^{-\sqrt{n}} e^{-t^2/2}\, dt$$

Write $z \equiv t + \sqrt{n}$. Then

$$\delta \int_0^\infty I(x)dx = \frac{e^{-n/2}}{\sqrt{n}} \int_{-\infty}^{0} e^{-z^2/2}\, e^{\sqrt{n}\,z}\, dz$$

This contribution is now bounded by

$$\left| \delta \int_0^\infty I(x)dx \right| \leq \frac{e^{-n/2}}{\sqrt{n}} \int_{-\infty}^{0} e^{\sqrt{n}z}\, dz$$
$$\leq \frac{e^{-n/2}}{n}$$

which is exponentially small.

The first approximation in the above retains just the first two terms in the Taylor series expansion of $\ln I(x)$ about $x = 1$. Suppose we retain the next two terms in that expansion. Use

$$\left[\frac{\partial^3}{\partial x^3} \ln I(x)\right]_{x=1} = 2n$$
$$\left[\frac{\partial^4}{\partial x^4} \ln I(x)\right]_{x=1} = -6n$$

Hence

$$\ln I(x) = -\frac{n}{2}(x-1)^2 + \frac{2n}{3!}(x-1)^3 - \frac{6n}{4!}(x-1)^4 + \cdots$$

Now go over to $t = \sqrt{n}\,(x-1)$, factor the exponential of the correction terms, and expand that exponential.[1] The correction to the integral is then

$$\delta \int_0^\infty I(x)dx = \frac{1}{\sqrt{n}} \int_{-\infty}^{\infty} dt\, e^{-t^2/2} \left[\frac{1}{3\sqrt{n}}\, t^3 - \frac{1}{4n} t^4 + \cdots\right]$$

[1]See Prob. 2.4(c) for a discussion of the correct limiting procedure to be used in computing these corrections.

The first correction is odd in t and integrates to zero. The second correction is explicitly of $O(1/n)$.[2]

In *summary*

$$\int_0^\infty I(x)dx = \left(\frac{2\pi}{n}\right)^{1/2}\left[1 + O\left(\frac{1}{n}\right)\right] \qquad ; \; n \to \infty$$

The correction is indeed of $O(1/n)$. This is Eq. (4.40).

(b) Equation (4.35) identifies the integral in part (a) as

$$\int_0^\infty x^{n-1}\, e^{n(1-x)}\, dx = \frac{e^n}{n^n}\Gamma(n)$$

Hence we have shown in (a) that

$$\frac{e^n}{n^n}\Gamma(n) = \left(\frac{2\pi}{n}\right)^{1/2}\left[1 + O\left(\frac{1}{n}\right)\right] \qquad ; \; n \to \infty$$

which is Eq. (4.41). If n is integer, then $\Gamma(n) = (n-1)!$, and with the neglect of the $O(1/n)$ corrections,

$$\ln{(n-1)!} = -n + n\ln n - \frac{1}{2}\ln n + O(1) \qquad ; \; n \to \infty \quad ; \text{ integer } n$$

$$\ln n! = n\ln n - n + \frac{1}{2}\ln n + O(1)$$

where we have added $\ln n$ to both sides to obtain the second line. This is Stirling's formula in Eq. (2.7)

$$\ln N! = N\ln N - N + O(\ln N) \qquad ; \text{ Stirling's formula}$$

(c) With the neglect of the $O(1/n)$ correction, the result in part (*b*) is

$$\frac{e^n}{n^n}\Gamma(n) = \left(\frac{2\pi}{n}\right)^{1/2}$$

[2]We can now investigate the term neglected in $\ln I(x)$ in Eq. (4.37) in the same fashion. From Eq. (4.36), with a Taylor series expansion about $x = 1$,

$$\delta \; \ln I(x) = -\ln x = -\left[(x-1) - \frac{1}{2}(x-1)^2 + \cdots\right]$$

Again go over to $t = \sqrt{n}\,(x-1)$, factor the exponential of the correction terms, and expand that exponential. The correction to the integral is then

$$\delta\int_0^\infty I(x)dx = \frac{1}{\sqrt{n}}\int_{-\infty}^\infty dt\, e^{-t^2/2}\left[-\frac{1}{\sqrt{n}}\,t + \frac{1}{2n}t^2 + \cdots\right]$$

The first correction is odd in t and integrates to zero; the second correction is again explicitly of $O(1/n)$.

Write this as

$$\Gamma(z) = \left(\frac{2\pi}{z}\right)^{1/2} \exp\left(z \ln z - z\right)$$

This is an analytic function of z in the cut z-plane, with a cut running along the negative z-axis from $-\infty$ to the origin. Hence this expression provides an analytic continuation of Stirling's formula to the complex z-plane.

Problem 4.2 Define the mean-square deviation from 1 of x in Fig. 4.3 in the text as

$$\langle (x-1)^2 \rangle \equiv \frac{\int_0^\infty (x-1)^2 I(x)dx}{\int_0^\infty I(x)dx}$$

where $I(x)$ is defined in Eq. (4.36). Take the root-mean-square deviation as a measure of the width of $I(x)$, and show that for large n.

$$\sqrt{\langle (x-1)^2 \rangle} = \frac{1}{\sqrt{n}} \qquad ; \; n \to \infty$$

Solution to Problem 4.2

The asymptotic form of the denominator is obtained in Eq. (4.40)

$$\begin{aligned} \int_0^\infty I(x)dx &\approx \int_0^\infty e^{-(x-1)^2 n/2}\, dx \\ &= \left(\frac{2\pi}{n}\right)^{1/2} \left[1 + O\left(\frac{1}{n}\right)\right] \qquad ; \; n \to \infty \end{aligned}$$

The corresponding asymptotic form of the numerator can be obtained by taking a derivative of this expression with respect to n. The leading term as $n \to \infty$ is

$$\begin{aligned} \int_0^\infty (x-1)^2 I(x)dx &\approx -2\frac{d}{dn} \int_0^\infty e^{-(x-1)^2 n/2}\, dx \\ &= -2\frac{d}{dn}\left(\frac{2\pi}{n}\right)^{1/2} \\ &= \frac{1}{n}\left(\frac{2\pi}{n}\right)^{1/2} \qquad ; \; n \to \infty \end{aligned}$$

The ratio then gives

$$\frac{\int_0^\infty (x-1)^2 I(x)dx}{\int_0^\infty I(x)dx} = \frac{1}{n} \qquad ; \; n \to \infty$$

Hence the root-mean-square deviation from 1 of x, which serves as a measure of the width of $I(x)$ in Fig. 4.3 in the text, is

$$\sqrt{\langle (x-1)^2 \rangle} = \frac{1}{\sqrt{n}} \qquad ; \; n \to \infty$$

Problem 4.3 Suppose the members of a given set of n_k systems in Fig. 2.1 in the text are in an excited state with energy ε_k and degeneracy ω_k. Show the number of complexions from this configuration is $\omega_k^{n_k}$.

Solution to Problem 4.3

Each of the n_k systems can be in any one of the ω_k states at the energy ε_k. The number of complexions for this system is then ω_k. The number of complexions from a configuration where the members of a given set of n_k systems in Fig. 2.1 in the text are in an excited state with energy ε_k and degeneracy ω_k is then

$$\begin{aligned} \Omega &= \omega_k \times \omega_k \times \omega_k \times \cdots \qquad ; \; n_k \text{ terms} \\ &= \omega_k^{n_k} \end{aligned}$$

Chapter 5

Applications of the Canonical Ensemble

Problem 5.1 Consider the triple sum over modes $\sum_{n_x} \sum_{n_y} \sum_{n_z} f(\mathbf{k}^2)$ where $k_i = (\pi/L)n_i$ with $i = (x, y, z)$, and $n_i = 1, 2, 3, \cdots$. Here f is an arbitrary function of $\mathbf{k}^2 = k_x^2 + k_y^2 + k_z^2$.

(a) Convert the sum to

$$\sum_{n_x} \Delta n_x \sum_{n_y} \Delta n_y \sum_{n_z} \Delta n_z \, f(\mathbf{k}^2)$$

where $\Delta n_i = 1$;

(b) Introduce $\Delta k_i \equiv (\pi/L)\Delta n_i$ and convert this sum to

$$\left(\frac{L}{\pi}\right)^3 \sum_{k_x} \Delta k_x \sum_{k_y} \Delta k_y \sum_{k_z} \Delta k_z \, f(\mathbf{k}^2)$$

(c) Now take the limit $L \to \infty$, and use the definition of the integral, to show

$$\sum_{n_x} \sum_{n_y} \sum_{n_z} f(\mathbf{k}^2) \to \left(\frac{L}{\pi}\right)^3 \int_{\text{first octant}} f(\mathbf{k}^2) \, d^3k \qquad ; \; L \to \infty$$

Solution to Problem 5.1

(a) We want the triple sum over all positive integers of

$$\mathcal{S} \equiv \sum_{n_x} \sum_{n_y} \sum_{n_z} f(\mathbf{k}^2) \qquad ; \; n_i = 1, 2, 3, \cdots$$

$$\mathbf{k}^2 = k_x^2 + k_y^2 + k_z^2 \qquad ; \; k_i = \frac{\pi}{L} n_i$$

Since $\Delta n_i = 1$, this is the same as

$$\mathcal{S} = \sum_{n_x} \Delta n_x \sum_{n_y} \Delta n_y \sum_{n_z} \Delta n_z \, f(\mathbf{k}^2) \qquad ; \ n_i = 1, 2, 3, \cdots$$

(b) Introduce $k_i = (\pi/L)n_i$, and observe that summing over k_i rather than n_i changes nothing.[1] With $\Delta k_i = (\pi/L)\Delta n_i$, the sum $\mathcal{S}$ then becomes

$$\mathcal{S} = \left(\frac{L}{\pi}\right)^3 \sum_{k_x} \Delta k_x \sum_{k_y} \Delta k_y \sum_{k_z} \Delta k_z \, f(\mathbf{k}^2) \qquad ; \ k_i = \frac{\pi}{L} n_i$$

(c) As $L \to \infty$ and the size of the box gets very large, the spacing between wavenumbers gets very small and $\Delta k_i = (\pi/L)\Delta n_i \to 0$. The sums then go over to integrals, and we have

$$\mathcal{S} \to \left(\frac{L}{\pi}\right)^3 \int_{\text{first octant}} f(\mathbf{k}^2) \, d^3k \qquad ; \ L \to \infty$$

Problem 5.2 (a) Start from Prob. 5.1(c), and show that

$$\sum_{n_x} \sum_{n_y} \sum_{n_z} f(\mathbf{k}^2) \to \left(\frac{L}{2\pi}\right)^3 \int_{\text{all k}} f(\mathbf{k}^2) \, d^3k \qquad ; \ L \to \infty$$

(b) Show it follows that

$$\sum_{n_x} \sum_{n_y} \sum_{n_z} f(\mathbf{k}^2) \to \left(\frac{L}{2\pi}\right)^3 \int_0^\infty f(k^2) \, 4\pi k^2 dk \qquad ; \ L \to \infty$$

Solution to Problem 5.2

(a) It is clear from Prob. 5.1(a) that the summand is an even function of each n_i. Hence instead of just summing over the positive integers, we can write each sum as 1/2 the sum over *all* integers

$$\mathcal{S} = \frac{1}{8} \sum_{n_x} \Delta n_x \sum_{n_y} \Delta n_y \sum_{n_z} \Delta n_z \, f(\mathbf{k}^2) \qquad ; \ n_i = \text{all integers}$$

$$; \ k_i = \frac{\pi}{L} n_i$$

[1] Note each $k_i \geq 0$; this is why the integral in (c) goes over the first octant.

Then, just as before,

$$\mathcal{S} = \left(\frac{L}{2\pi}\right)^3 \sum_{k_x} \Delta k_x \sum_{k_y} \Delta k_y \sum_{k_z} \Delta k_z \, f(\mathbf{k}^2) \qquad ; \; k_i = \frac{\pi}{L} n_i$$

where the sum now goes over *all* $k_i = (\pi/L)n_i$. It follows that

$$\mathcal{S} \to \left(\frac{L}{2\pi}\right)^3 \int_{\text{all k}} f(\mathbf{k}^2)\, d^3k \qquad ; \; L \to \infty$$

The integral over the first octant is 1/8 the integral over all k.

(b) The integrand is assumed to be only a function of $\mathbf{k}^2$. Hence the integral over angles can be performed immediately, with the result

$$\mathcal{S} \to \left(\frac{L}{2\pi}\right)^3 \int_0^\infty f(k^2)\, 4\pi k^2 dk \qquad ; \; L \to \infty$$

This is the desired answer.

Problem 5.3 A Debye temperature of $\theta_D = 1890^\circ$K is found to give an excellent fit to the molar specific heat of diamond (see Fig. 5.5 in the text). Diamond is pure carbon with an atomic weight of 12 (so that 12 gm = 1 mole), and the measured mass density of diamond is approximately $3.25\,\text{gm/cm}^3$.

(a) Express c_{av} in terms of θ_D in the Debye model, and deduce the speed of sound in diamond;

(b) Compare the value of c_{av} found in part (a) with the measured speed of sound in diamond.[2]

Solution to Problem 5.3

(a) From Eqs. (5.37) and (5.31), the relation between the cut-off frequency ν_m and the average speed of sound c_{av} in the Debye model is

$$\nu_m = \left[\frac{3}{4\pi}\left(\frac{N}{V}\right)\right]^{1/3} c_{\text{av}} \qquad ; \; \theta_D = \frac{h\nu_m}{k_\text{B}}$$

It follows that

$$\frac{c_{\text{av}}}{c} = \frac{k_\text{B}\theta_D}{2\pi\hbar c}\left(\frac{4\pi}{3n}\right)^{1/3}$$

where $n = N/V$ is the number density, and c is the speed of light.

[2]One goal of this part of the problem is to get the reader to locate such measured values. (*Hint*: Try the *Handbook of Chemistry and Physics*, or the *Web*.)

We are given that diamond has a mass density $\rho = 3.25\,\text{gm/cm}^3$ and carbon has 12 gm/mole. Use Avagadro's constant

$$N_{\text{A}} = 6.022 \times 10^{23}\,/\text{mole}$$

This gives[3]

$$\left(\frac{3n}{4\pi}\right)^{1/3} = 3.39 \times 10^7\,/\text{cm}$$

Now use the additional physical constants

$$\begin{aligned} k_{\text{B}} &= 8.620 \times 10^{-5}\,\text{eV/}^{\text{o}}\text{K} \\ \hbar c &= 1.973 \times 10^{-5}\,\text{eV-cm} \\ c &= 2.998 \times 10^{10}\,\text{cm/sec} \end{aligned}$$

Then, with the given value for diamond of $\theta_D = 1890\,^{\text{o}}\text{K}$,

$$c_{\text{av}} = 1.16 \times 10^6\,\text{cm/sec}$$

(b) The experimental speed of sound in diamond is[4]

$$c_{\text{sound}} = 1.20 \times 10^6\,\text{cm/sec} \qquad ; \text{ experiment}$$

Problem 5.4 (a) Show that the minimum wavelength in the Debye model is

$$\lambda_{\text{min}} = \frac{c_{\text{av}}}{\nu_m} = \left[\frac{3}{4\pi}\left(\frac{N}{V}\right)\right]^{-1/3}$$

(b) Evaluate this quantity for diamond (see Prob. 5.3). Discuss.

Solution to Problem 5.4

(a) In the Debye model, the minimum wavelength λ_{min} for a sound wave can be related to the maximum frequency ν_m by

$$\lambda_{\text{min}}\, \nu_m = c_{\text{av}}$$

[3]Note $n/N_A = (3.25\,\text{gm/cm}^3)/(12\,\text{gm/mole})$.

[4]From the *Web*.

It then follows from Prob. 5.3(a) that the relation between λ_{min} and the density n is

$$\lambda_{\text{min}} \equiv \frac{c_{\text{av}}}{\nu_m} = \left(\frac{4\pi}{3n}\right)^{1/3}$$

(b) Again, from Prob. 5.3(a), for diamond

$$\left(\frac{4\pi}{3n}\right)^{1/3} = 2.95 \times 10^{-8}\,\text{cm}$$

One cannot have a normal mode of oscillation in a solid with wavelength shorter than the interparticle spacing. Thus one immediately has

$$\lambda \geq n^{-1/3}$$

This is reflected in the above result for λ_{min}. The underlying lattice provides a natural cutoff to the normal-mode spectrum in a solid that is absent for the normal modes of radiation in a cavity (compare Prob. 7.3).

Problem 5.5 Start from the spectral weight in the lattice model in Eq. (5.65), and verify that the total number of normal modes is given by $\int_0^{\nu_M} g(\nu)d\nu = N$.

Solution to Problem 5.5 The spectral weight in Eq. (5.65) is

$$g(\nu) = \frac{2N}{\pi(\nu_M^2 - \nu^2)^{1/2}} \qquad ; \text{ spectral weight}$$

Consider the sum over modes

$$\int_0^{\nu_M} g(\nu)\,d\nu = \frac{2N}{\pi}\int_0^{\nu_M} \frac{d\nu}{(\nu_M^2 - \nu^2)^{1/2}}$$

Change variables to $\nu/\nu_M = x$

$$\int_0^{\nu_M} g(\nu)\,d\nu = \frac{2N}{\pi}\int_0^1 \frac{dx}{(1-x^2)^{1/2}} \qquad ; \nu/\nu_M = x$$

Now introduce $x = \sin\theta$ with $dx = \cos\theta\,d\theta$

$$\begin{aligned}\int_0^{\nu_M} g(\nu)\,d\nu &= \frac{2N}{\pi}\int_0^{\pi/2} d\theta \qquad ; x = \sin\theta \\ &= N\end{aligned}$$

This is the given result.

Problem 5.6 (a) Do the integral numerically, and make a good plot of the molar heat capacity in the Debye model in Eq. (5.33). Plot the result as a function of T/θ_D;

(b) Plot the molar heat capacity in the Einstein model in Eq. (2.53) as a function of T/θ_E, where $\theta_E \equiv h\nu_0/k_B$ (here ν_0 denotes the single oscillator frequency). Compare with the result in (a).

Solution to Problem 5.6

(a) The molar heat capacity in the Debye model is given in Eq. (5.33)

$$\frac{C_V}{R} = 9\left(\frac{T}{\theta_D}\right)^3 \int_0^{\theta_D/T} \frac{u^4 e^u\, du}{(e^u - 1)^2} \qquad ; \text{ Debye}$$

where $\theta_D = h\nu_m/k_B$.

(b) The molar heat capacity in the Einstein model is given in Eq. (2.53)

$$\frac{C_V}{R} = 3\left(\frac{\theta_E}{T}\right)^2 \frac{e^{\theta_E/T}}{(e^{\theta_E/T} - 1)^2} \qquad ; \text{ Einstein}$$

where $\theta_E = h\nu_0/k_B$.

The molar heat capacities in the Debye model and the Einstein model, calculated with Mathcad 7, are plotted in Fig. 5.1.

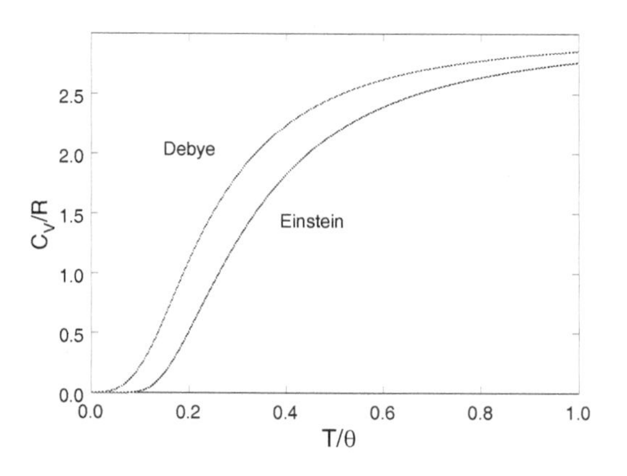

Fig. 5.1 Molar heat capacities in the Debye model ($\theta_D = h\nu_m/k_B$) and Einstein model ($\theta_E = h\nu_0/k_B$).

While both heat capacities go to the same values at high T

$$\frac{C_V}{R} \to 3 \qquad\qquad ; \; T \to \infty$$

their low-T behavior is quite distinct

$$\begin{aligned} \frac{C_V}{R} &\to \frac{12\pi^4}{5}\left(\frac{T}{\theta_D}\right)^3 && ; \text{ Debye} \qquad ; \; T \to 0 \\ &\to 3\left(\frac{\theta_E}{T}\right)^2 e^{-\theta_E/T} && ; \text{ Einstein} \end{aligned}$$

Problem 5.7 Why is it inappropriate to use $\kappa\{\varepsilon[x+u(x)+\Delta x] - \varepsilon[x+u(x)]\}$ in Eq. (5.39)?

Solution to Problem 5.7

Suppose we use this, which from Fig. 5.8 in the text would seem to be correct. Then for small Δx, Eq. (5.39) would read

$$F_{\text{el}} = \kappa \frac{\partial^2 u(x+u)}{\partial x^2} \Delta x$$

For small displacements $u(x)$, this becomes

$$F_{\text{el}} \approx \kappa \left\{ \frac{\partial^2}{\partial x^2}\left[u(x) + u(x)\frac{\partial u(x)}{\partial x}\right]\right\} \Delta x$$

The second term in square brackets represents a correction to the wave equation that is *non-linear* in $u(x)$, and, in the end, we assume the displacement $u(x)$ is small and seek the *linear* wave equation.

Problem 5.8

Consider the transverse planar oscillations of a string with displacement $q(x,t)$ and fixed endpoints. The normal-mode amplitudes are $q(x) = A\sin kx$, with $k = n\pi/L$, $n = 1, 2, 3, \cdots$, and $\omega = kc$. Let $L \to \infty$.

(a) Show the spectral weight is $(1/L)g(\nu) = 2/c$;

(b) Show the normal-mode amplitude is the sum of two waves, one running to the right and one to the left;

(c) Shift the phase of one of the running waves by $\delta(k)$. Show the values of q on the far-away boundaries are altered, while the result in part (a) remains unchanged.

Solution to Problem 5.8

(a) The normal-mode amplitudes for the transverse oscillations of a string of length L with fixed endpoints

$$q(0,t) = q(L,t) = 0 \qquad ; \text{ fixed endpoints}$$

are

$$q(x,t) = \left(\frac{2}{L}\right)^{1/2} \sin(kx)\cos(\omega_k t) \qquad ; \; k_n = \frac{\pi n}{L} \qquad ; \; n = 1, 2, \cdots$$
$$; \; \omega_k = kc$$

where c is the velocity of sound in the string. These are *standing waves*.

For large L, the sum over modes is converted to an integral as in Probs. 5.1–5.2

$$\sum_n f(k_n^2) \to \frac{L}{\pi}\int_0^\infty dk\, f(k^2) \qquad ; \; L \to \infty$$

The spectral weight is then

$$\frac{1}{L}g(\nu)\,d\nu = \frac{1}{\pi}dk = \frac{1}{\pi c}d\omega = \frac{2}{c}d\nu$$

where $\omega = 2\pi\nu$. Hence

$$\frac{1}{L}g(\nu) = \frac{2}{c}$$

(b) Note that

$$\sin(kx) = \frac{1}{2i}\left[e^{ikx} - e^{-ikx}\right]$$

These yield *running waves*, with waves going to the right and to the left. We could just as well have joined the two ends of the string around a cylinder and introduced *periodic boundary conditions*

$$q(x,t) = q(x+L,t) \qquad ; \text{ periodic boundary conditions}$$

The normal modes in this case are

$$q(x,t) = \frac{1}{\sqrt{L}}e^{i(kx-\omega_k t)} \qquad ; \; k_n = \frac{2\pi n}{L} \qquad ; \; n = 0, \pm 1, \pm 2, \cdots$$
$$; \; \omega_k = |k|c$$

For large L, the sum over modes is converted to an integral as

$$\sum_n f(k_n^2) \to \frac{L}{2\pi}\int_{-\infty}^{\infty} dk\, f(k^2)$$
$$= \frac{L}{\pi}\int_0^{\infty} dk\, f(k^2)$$

Hence the spectral weight is again

$$\frac{1}{L}g(\nu) = \frac{2}{c}$$

(c) If we were to now shift the phase of those waves running, say, to the left in part (b) by $\delta(k)$, we would not affect the counting of modes, and hence the spectral density would be unaffected.

The local spectral density $g(\nu)/L$ is *independent* of the conditions on the far-away boundaries.

Problem 5.9 Consider a mass m attached to one end of a massless spring, with the other end fixed, and take $x = 0$ to mark the unextended length of the spring. Apply a constant force F to the mass which stretches the spring. The hamiltonian for this one-dimensional system is $H = p^2/2m + \kappa x^2/2 - Fx$.[5]

(a) Complete the square, and show the classical partition function is $(\text{p.f.}) = (\text{p.f.})_{\text{osc}}\, e^{F^2/2\kappa k_{\text{B}}T}$ where $(\text{p.f.})_{\text{osc}}$ is the result for a free oscillator with $F = 0$;

(b) Show the Helmholtz free energy of an assembly of N such springs is $A(N,T,V) = A(N,T,V)_{\text{osc}} - NF^2/2\kappa$;

(c) Show the change in the internal energy from the assembly at $F = 0$ is $\Delta E = -NF^2/2\kappa$; Derive this result by balancing forces in a system. What is the corresponding entropy change ΔS?

(d) Increase F slighty so that $F \to F + dF$. Show that the work done *by* a system is $dW = F\,dF/\kappa$. Explain the sign. Hence show that the work done by N systems when the force is increased from 0 to F is $\Delta W = NF^2/2\kappa$;

(f) Show that the isothermal heat flow when the force is increased reversibly from 0 to F, is given by $\Delta Q = T\Delta S$, where ΔS is the entropy change calculated in (c).

[5]An example is provided by gravity, with the mass hanging down on the spring and $F = mg$; one could then change F by changing g. It helps in visualizing this problem to then put an imaginary box around the whole system, including the mass.

Solution to Problem 5.9

If we accept the example, and measure x downward, then the gravitational potential per unit mass is

$$\phi_g = -gx$$

and the hamiltonian is

$$H = H_0 + m\phi_g = \frac{p^2}{2m} + \frac{1}{2}\kappa x^2 - mgx$$

Thus, let us consider the generic problem

$$H = \frac{p^2}{2m} + \frac{1}{2}\kappa x^2 - Fx$$

(a) The classical partition function is

$$(\text{p.f.})_{\text{F}} = \frac{1}{h}\int_{-\infty}^{\infty} e^{-p^2/2mk_{\text{B}}T}\,dp \int_{-\infty}^{\infty} e^{-(\kappa x^2-2Fx)/2k_{\text{B}}T}\,dx$$

Complete the square in the second exponent

$$(\text{p.f.})_{\text{F}} = \frac{1}{h}e^{F^2/2\kappa k_{\text{B}}T}\int_{-\infty}^{\infty} e^{-p^2/2mk_{\text{B}}T}\,dp \int_{-\infty}^{\infty} e^{-\kappa(x-F/\kappa)^2/2k_{\text{B}}T}\,dx$$

Now change variable in the spatial integral $x \to x - F/\kappa$, to obtain

$$(\text{p.f.})_{\text{F}} = e^{F^2/2\kappa k_{\text{B}}T}(\text{p.f.})_{\text{osc}}$$

where $(\text{p.f.})_{\text{osc}}$ is the result for a free oscillator with $F = 0$.

(b) The Helmholtz free energy of N such systems is

$$\begin{aligned} A(N,T,V,F) &= -Nk_{\text{B}}T\ln(\text{p.f.})_{\text{F}} \\ &= A(N,T,V)_{\text{osc}} - N\frac{F^2}{2\kappa} \end{aligned}$$

(c) The internal energy is obtained from the Helmholtz free energy through Eq. (2.48)

$$E = -T^2\frac{\partial}{\partial T}\left(\frac{A}{T}\right)_V$$

Hence, the change in the internal energy from the assembly at $F = 0$ is

$$\Delta E = -N\frac{F^2}{2\kappa}$$

The entropy is obtained from the Helmholtz free energy through Eq. (1.20)

$$S = -\left(\frac{\partial A}{\partial T}\right)_V$$

Therefore, the corresponding change in the entropy from the assembly at $F = 0$ is

$$\Delta S = 0$$

(d) Suppose we balance forces on the mass in the example

$$\kappa x = mg \qquad ; \text{ balance forces}$$

We can now understand the sign of ΔE in (c). The decrease in potential energy for a system is $-mgx$, while the increase in energy stored in the spring is $\kappa x^2/2$. Hence

$$\Delta E = N\left[\frac{1}{2}\kappa x^2 - mgx\right] = -N\frac{(mg)^2}{2\kappa} = -N\frac{F^2}{2\kappa}$$

Because of the decrease in gravitational potential energy, the energy of the combined system is less than when the mass sits at the origin, and the spring is unextended.

(e) Make an isothermal, quasistatic increase in the gravitational field $g \to g + dg$. Then the mass moves to $x + dx$ where

$$\kappa\, dx = m\, dg$$

As the strength of the gravitational field increases, the mass will fall and the falling mass can be used to produce external work. How much external work can be produced? The infinitesimal amount of gravitational work available is[6]

$$dw = F_g\, dx = mg\, dx = \frac{m^2}{\kappa} g\, dg$$

The total amount of gravitational work available for external use is now obtained by integrating from 0 to g

$$w = \int_0^g \frac{m^2}{\kappa} g\, dg = \frac{(mg)^2}{2\kappa}$$

[6]The work done against the spring is stored in the potential energy of the spring. There is an alternate interpretation of the relation $dw = (mg/\kappa)d(mg)$. An infinitesimal, incremental element of weight $d(mg)$ falls a distance $x = mg/\kappa$ and is available to do external work; the total amount of external work is then obtained by integration.

Generically, for N systems, this work is

$$W = N\frac{F^2}{2\kappa}$$

(e) We observe that the external work done is equal to the decrease in internal energy

$$\Delta E = -W$$

Hence, by the first law, there is no reversible heat flow

$$Q = 0$$

Therefore, the entropy does not change during this process.

$$T\Delta S = 0$$

This reproduces the result in part (c).

Problem 5.10 (a) Start from the partition function (p.f.) in Prob. 5.9(a). Show the mean displacement per system is $\langle x\rangle = k_{\rm B}T(\partial/\partial F)\ln{\rm (p.f.)} = F/\kappa$;

(b) Similarly, show $\langle\varepsilon\rangle = k_{\rm B}T^2(\partial/\partial T)\ln{\rm (p.f.)} = \langle\varepsilon\rangle_{\rm osc} - F^2/2\kappa$;

(c) Write the Helmholtz free energy per system as $a(T,V,N,F) = -k_{\rm B}T\ln{\rm (p.f.)}$ and work per system as dw. Show

$$\left(\frac{\partial a}{\partial F}\right)_{T,V,N} dF = -\left(\frac{F}{\kappa}\right) dF = -\langle x\rangle\, dF = -dw$$

where dw follows from Prob. 5.9(d). Hence write the analog of Eq. (3.180).

Solution to Problem 5.10

(a) The mean displacement $\langle x\rangle$ is calculated from the partition function in Prob. 5.9(a) as

$$\begin{aligned}\langle x\rangle &= \frac{1}{{\rm (p.f.)_F}}\frac{1}{h}\int_{-\infty}^{\infty} e^{-p^2/2mk_{\rm B}T}\,dp\int_{-\infty}^{\infty} x\,e^{-(\kappa x^2-2Fx)/2k_{\rm B}T}\,dx\\ &= k_{\rm B}T\frac{\partial}{\partial F}\ln{\rm (p.f.)_F}\end{aligned}$$

Here all other variables in the set (N,T,V,F) are to be held fixed in taking the partial derivative. Since the partition function is

$${\rm (p.f.)_F} = e^{F^2/2\kappa k_{\rm B}T}\,{\rm (p.f.)_{osc}}$$

we find

$$\langle x \rangle = \frac{F}{\kappa}$$

(b) In a similar fashion

$$\begin{aligned}\langle \varepsilon \rangle &= \frac{1}{(\text{p.f.})_{\text{F}}} \frac{1}{h} \int_{-\infty}^{\infty} e^{-p^2/2mk_{\text{B}}T}\, dp \int_{-\infty}^{\infty} H\, e^{-(\kappa x^2 - 2Fx)/2k_{\text{B}}T}\, dx \\ &= k_{\text{B}}T^2 \frac{\partial}{\partial T} \ln(\text{p.f.})_{\text{F}}\end{aligned}$$

which gives

$$\langle \varepsilon \rangle = \langle \varepsilon \rangle_{\text{osc}} - \frac{F^2}{2\kappa}$$

(c) Write the Helmholtz free energy per system in Prob. 5.9 as

$$\begin{aligned}a(T, V, N, F) &\equiv -k_{\text{B}}T \ln(\text{p.f.})_{\text{F}} \\ &= a(T, V, N)_{\text{osc}} - \frac{F^2}{2\kappa}\end{aligned}$$

The differential external work per system dw from Prob. 5.9(e) is

$$dw = \left(\frac{m^2 g}{\kappa}\right) dg = \left(\frac{F}{\kappa}\right) dF$$

Hence, we have

$$\frac{\partial a(T, V, N, F)}{\partial F}\, dF = -\left(\frac{F}{\kappa}\right) dF = -\langle x \rangle\, dF = -dw$$

The analog of the thermodynamic relation in Eq. (3.180) therefore reads

$$dA = -SdT - PdV + \mu dN - \mathcal{D}dF$$

Where $\mathcal{D}$ is the total induced dipole moment

$$\mathcal{D} = N\langle x \rangle$$

Problem 5.11 The Lennard-Jones (or "6-12") interatomic two-body potential is a useful empirical potential of the form $U(r) = 4\epsilon[(\sigma/r)^{12} - (\sigma/r)^6]$ where (ϵ, σ) are positive constants (see Fig. 5.19 in the text). Calculate the second virial coefficient $B(T)$ at high temperature with this potential.

Solution to Problem 5.11

The Lennard-Jones potential is

$$U(r) = 4\epsilon \left[\left(\frac{\sigma}{r}\right)^{12} - \left(\frac{\sigma}{r}\right)^{6} \right]$$

The second virial coefficient is given by Eq. (5.88)

$$B(T) = 2\pi \int_0^\infty \left[1 - e^{-U(r)/k_{\rm B}T}\right] r^2 dr$$

This integral is well-defined since $e^{-U(r)/k_{\rm B}T}$ vanishes as $r \to 0$ [compare Fig. (5.13) in the text]; however, the exponential cannot be expanded at high T, since the integral over $U(r)$ itself does not converge at the origin.[7] Thus we resort to numerical integration of $B(T)$. The result obtained with Mathcad 7 is shown in Fig. 5.2.[8]

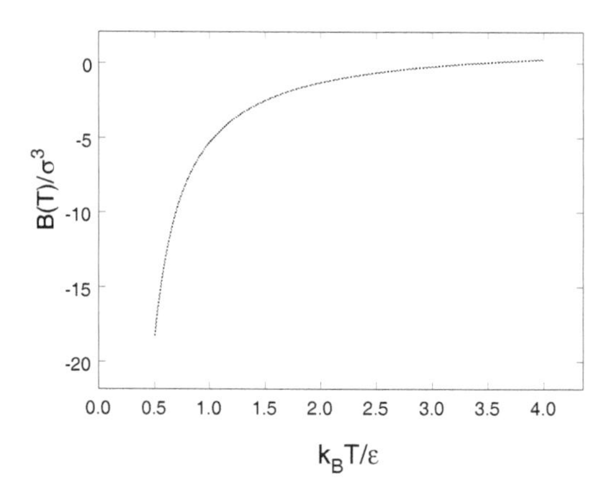

Fig. 5.2 The second virial coefficient $B(T)/\sigma^3$ for the Lennard-Jones potential obtained by numerical integration is plotted against $k_{\rm B}T/\epsilon$. The upper and lower limits on the radial integral are finite and varied until the result is stable.

Problem 5.12 The model interatomic two-body potential in Eq. (5.89) has a hard core at a distance σ in the relative coordinate r, surrounded by a $-c/r^6$ Van der Waal's attraction.

---

[7] The exponential function has an *essential singularity* at the origin; this problem turns out to be more difficult than intended!

[8] We use the program Axum 6.0 LE, distributed with Mathcad, to plot the Mathcad results.

(a) Identify Van de Waal's parameters (a, b) in terms of the parameters (σ, c) by matching the second virial coefficients in Eqs. (5.92) and (5.75);[9]

(b) Discuss the role of the parameters (a, b) in Van de Waal's equation of state in Eq. (5.73) in terms of the properties of the potential.

Solution to Problem 5.12 (a) The Van der Waal's potential in Eq. (5.89) is[10]

$$\begin{aligned} U(r) &= \infty & ; r < \sigma \\ &= -\frac{c}{r^6} & ; r > \sigma \end{aligned}$$

The corresponding high-temperature second virial coefficient is calculated in Eq. (5.92)

$$B(T) = \frac{2\pi\sigma^3}{3}\left[1 - \frac{c}{\sigma^6 k_{\mathrm{B}}T}\right] \qquad ; T \to \infty$$

The Van der Waal's equation of state is re-written in Eq. (5.136) as

$$P(v) = \frac{k_{\mathrm{B}}T}{v - b} - \frac{a}{v^2} \qquad ; \text{Van der Waal's}$$

The second virial coefficient for this equation of state is [see Eq. (5.75)]

$$B(T)_{\mathrm{VW}} = b - \frac{a}{k_{\mathrm{B}}T}$$

A comparison then allows us to express the parameters (a, b) in the Van der Waal's equation of state in terms of the parameters (σ, c) in the potential

$$b = \frac{2\pi\sigma^3}{3} \qquad ; a = \frac{2\pi c}{3\sigma^3}$$

(b) The qualitative arguments for the two modifications of the perfect gas law $P = k_{\mathrm{B}}T/v$ in the Van der Waal's equation of state are

- An attractive interaction between two systems in the gas, characterized by a positive constant a, will reduce the pressure $P(v)$;
- The size of the systems, characterized by b, will decrease the available volume per particle v.

---

[9] Equation (5.75) corrects a misprint in Eq. (5.93).

[10] See Fig. 5.15 in the text; note that in this potential c denotes the strength of the attractive tail, and *not* the speed of light.

Here we observe that the constant b is indeed four times the hard sphere volume in the potential (see Prob. 5.15), and the constant a is, in turn, proportional to the strength c of the attractive tail of the two-body interaction. Furthermore, the volume and temperature dependences of the two *ad hoc* (a, b) contributions in the Van der Waal's equation of state are here justified in terms of an underlying two-body potential.

Problem 5.13 (a) Start from Eq. (5.97) and explicitly derive the linked-cluster decomposition in Eqs. (5.108) for $N = 4$;

(b) Show that the expression for the cluster of 3 systems in Eq. (5.101) is symmetric under the interchange of any pair of indices (i, j, k).

Solution to Problem 5.13

(a) We start from the configuration integral in Eq. (5.97)

$$Q_N = \int \cdots \int d\tau_1 \cdots d\tau_N \prod_{i<j\leq N} (1 + f_{ij})$$

The particles are now all labeled, and the product goes over all pairs of particles. Suppose that there are four systems, and $N = 4$. Then

$$\prod_{i<j\leq 4} (1 + f_{ij}) = (1 + f_{12})(1 + f_{13})(1 + f_{14})(1 + f_{23})(1 + f_{24})(1 + f_{34})$$

The terms in this repeated product are now partitioned into a number (n_1, n_2, n_3, n_4) of linked clusters where the partitions satisfy Eq. (5.100)

$$\sum_{l=1}^{4} l n_l = 4$$

The number of such partitions $g(n_1, n_2, n_3, n_4)$ is given in Eq. (5.107)

$$g(n_1, n_2, n_3, n_4) = \frac{4!}{n_1!n_2!n_3!n_4!(1!)^{n_1}(2!)^{n_2}(3!)^{n_3}(4!)^{n_4}}$$

The possible partitions (n_1, n_2, n_3, n_4), and the number of each type, are

$$\begin{aligned}
&(4,0,0,0) \qquad && ;\ g(4,0,0,0) = 1 \\
&(2,1,0,0) && ;\ g(2,1,0,0) = 6 \\
&(0,2,0,0) && ;\ g(0,2,0,0) = 3 \\
&(1,0,1,0) && ;\ g(1,0,1,0) = 4 \\
&(0,0,0,1) && ;\ g(0,0,0,1) = 1
\end{aligned}$$

The nature of the linked clusters is as follows:

(1) The first partition represents just four isolated systems. It is the 1 term in the repeated product;
(2) The second partition represents one linked pair and two isolated systems. There are 6 such terms in the repeated product

$$f_{12} + f_{13} + f_{14} + f_{23} + f_{24} + f_{34}$$

(3) The third partition represents two linked pairs. There are 3 such terms in the repeated product

$$f_{12}f_{34} + f_{13}f_{24} + f_{14}f_{23}$$

(4) The fourth term represents a linked cluster of three systems and one isolated system. The cluster is illustrated graphically in Fig. 5.16 in the text. There are four such clusters, corresponding to the omission of each of the four particles in turn, and each such cluster constitutes four terms in the repeated product, for a total of 16 terms;
(5) There is one cluster composed of all four linked systems. It is illustrated in Fig. 5.3, and it is composed of 38 terms in the repeated product.

There are $2^6 = 64$ terms in the repeated product, and the above enumeration accounts for all of them.

Fig. 5.3 Linked clusters of four labeled systems sorted according to the number of open links on the perimeter.

Given the definition of the cluster integrals in Eqs. (5.102)–(5.105), this analysis reduces the above configuration integral to Eq. (5.106)

$$Q_4 = \sum_{\{n_l\}} g(n_1, n_2, n_3, n_4)(VB_1)^{n_1}(VB_2)^{n_2}(VB_3)^{n_3}(VB_4)^{n_4}$$

Hence, in summary, for $N = 4$ systems we are led to the following relations [see Eqs. (5.108)]

$$\frac{Q_4}{4!} = \sum_{\{n_l\}} \frac{1}{n_1!n_2!n_3!n_4!}\left(\frac{VB_1}{1!}\right)^{n_1}\left(\frac{VB_2}{2!}\right)^{n_2}\left(\frac{VB_3}{3!}\right)^{n_3}\left(\frac{VB_4}{4!}\right)^{n_4}$$

$$N = \sum_{l=1}^{4} ln_l$$

The sum in Q_4 goes over the five partitions $\{n_l\} = (n_1, n_2, n_3, n_4)$ satistfying the second relation, and the cluster integrals are given in Eq. (5.104). *The above equations form an exact linked-cluster decomposition of the configuration integral for* $N = 4$.

(b) Suppose the (labeled) systems $(i, j, k) = (1, 2, 3)$ form a cluster of 3 systems. The contribution of this three-body cluster is illustrated in Fig. 5.16 in the text, and it takes the form in Eq. (5.101)

$$F(1, 2, 3) \equiv f_{12}f_{23} + f_{13}f_{23} + f_{12}f_{13} + f_{12}f_{23}f_{13}$$

The coordinates are all equivalent. Let us interchange, say, $1 \rightleftharpoons 2$. Then[11]

$$\begin{aligned} F(2, 1, 3) &= f_{12}f_{13} + f_{23}f_{13} + f_{12}f_{23} + f_{12}f_{13}f_{23} \\ &= F(1, 2, 3) \end{aligned}$$

Hence, $F(1, 2, 3)$ is symmetric under the interchange $1 \rightleftharpoons 2$. Since the coordinates are all equivalent, this holds for the interchange of any pair.[12]

Problem 5.14 This problem uses Euler's theorem on homogeneous functions to prove that the Gibbs free energy can be expressed as $G = N\mu$ where μ is the chemical potential.

(a) Give the argument that $G(T, P, \alpha N) = \alpha G(T, P, N)$;

(b) Differentiate with respect to α, and then set $\alpha = 1$, to show

$$N\left(\frac{\partial G}{\partial N}\right)_{T,P} = N\mu = G(T, P, N)$$

This is an *important result.*

[11]Recall $f_{ij} = f_{ji}$ is symmetric.

[12]Try it!

Solution to Problem 5.14

(a) The Gibbs free energy $G(T, P, N)$ is *extensive*, and scales with the number of systems. The thermodynamic variables (T, P) are both *intensive*, independent of the size of the assembly. Thus we immediately arrive at the scaling relation

$$G(T, P, \alpha N) = \alpha G(T, P, N)$$

(b) Differentiate this relation with respect to α

$$\frac{d}{d\alpha} G(T, P, \alpha N) = N \frac{\partial}{\partial(\alpha N)} G(T, P, \alpha N) = G(T, P, N)$$

Now set $\alpha = 1$

$$N \frac{\partial}{\partial N} G(T, P, N) = G(T, P, N)$$

From Eq. (1.27), the partial derivative of the Gibbs free energy with respect to the number of systems at fixed (T, P) is just the chemical potential

$$\frac{\partial}{\partial N} G(T, P, N) = \mu$$

Hence we arrive at the important relation

$$G(T, P, N) = N\mu$$

Problem 5.15 Consider the hard-sphere gas where there is a hard-core in the two-body potential that extends out to a distance σ in the relative coordinate. Show the second virial coefficient is

$$B(T) = \frac{2\pi\sigma^3}{3} = 4v_0 \qquad ; \text{ hard sphere}$$

where v_0 is the volume of each hard sphere.

Solution to Problem 5.15

The second virial coefficient is defined in Eq. (5.88)

$$B(T) = 2\pi \int_0^\infty \left[1 - e^{-U(r)/k_B T}\right] r^2 dr$$

Consider the hard-core potential

$$\begin{aligned} U(r) &= \infty \qquad &; r \leq \sigma \\ &= 0 \qquad &; r > \sigma \end{aligned}$$

where r is the relative coordinate of two systems. Then (compare Fig. 5.13 in the text)

$$\begin{aligned}\left[1 - e^{-U(r)/k_{\rm B}T}\right] &= 1 \qquad &; r \leq \sigma \\ &= 0 \qquad &; r > \sigma\end{aligned}$$

It follows that

$$B(T) = 2\pi \int_0^\sigma r^2 dr = \frac{2\pi\sigma^3}{3}$$

Since the radius of each hard sphere is $\sigma/2$, and its volume $v_0 = \pi\sigma^3/6$, this result for the second virial coefficient is

$$B(T) = 4v_0$$

where v_0 is the volume of each hard sphere.

Problem 5.16 Show the third virial coefficient with the Van der Waal's equation of state is

$$C(T) = b^2 \qquad ; \text{Van der Waal's}$$

Solution to Problem 5.16

The virial expansion is defined in Eq. (5.72)

$$\frac{P}{k_{\rm B}T} = \sum_{n=1}^{\infty} \frac{c_n(T)}{v^n} = \frac{1}{v} + \frac{B(T)}{v^2} + \frac{C(T)}{v^3} + \cdots$$

The Van der Waal's equation of state is given in Eq. (5.136)

$$\begin{aligned}\left(P + \frac{a}{v^2}\right)(v - b) &= k_{\rm B}T \qquad ; \text{Van der Waal's} \\ P(v) &= \frac{k_{\rm B}T}{v - b} - \frac{a}{v^2}\end{aligned}$$

The last equation is manipulated to

$$\begin{aligned}\frac{P(v)}{k_{\rm B}T} &= \frac{1}{v - b} - \left(\frac{a}{k_{\rm B}T}\right)\frac{1}{v^2} \\ &= \frac{1}{v} + \left(b - \frac{a}{k_{\rm B}T}\right)\frac{1}{v^2} + \frac{b^2}{v^3} + \cdots\end{aligned}$$

Hence we can identify the second and third virial coefficients of the Van der Waal's gas

$$\begin{aligned} B(T) &= b - \frac{a}{k_{\rm B}T} \qquad ; \text{ Van der Waal's} \\ C(T) &= b^2 \end{aligned}$$

The result for $B(T)$ is just that of Eq. (5.75), and $C(T)$ is as given.

Problem 5.17 (a) Show that the third virial coefficient of a real gas is given by

$$C(T) = \frac{4g_2^2 - 2g_3g_1}{g_1^4}$$

(b) Show this reduces to

$$C(T) = B_2^2 - \frac{1}{3}B_3 = -\frac{1}{3V}\int\int\int d\tau_1 d\tau_2 d\tau_3\, f_{12}f_{23}f_{31}$$

Solution to Problem 5.17

(a) We are instructed to retain the cubic terms in the parametric equation of state in Eqs. (5.126) and (5.128)

$$\begin{aligned} \frac{1}{v} &= g_1\lambda + 2g_2\lambda^2 + 3g_3\lambda^3 + \cdots \\ \frac{P}{k_{\rm B}T} &= g_1\lambda + g_2\lambda^2 + g_3\lambda^3 + \cdots \end{aligned}$$

Expand λ in inverse powers of $1/v$ according to

$$\lambda = \frac{a_1}{v} + \frac{a_2}{v^2} + \frac{a_3}{v^3}\cdots$$

Now determine the coefficients (a_1, a_2, a_3) by substituting this in the first equation above, just as in Eqs. (5.129)

$$\frac{1}{v} = g_1\left(\frac{a_1}{v} + \frac{a_2}{v^2} + \frac{a_3}{v^3}\right) + 2g_2\left(\frac{a_1^2}{v^2} + \frac{2a_1a_2}{v^3}\right) + 3g_3\left(\frac{a_1^3}{v^3}\right) + \cdots$$

The inversion of the power series is thus obtained through

$$\begin{aligned} g_1a_1 &= 1 \\ g_1a_2 + 2g_2a_1^2 &= 0 \\ g_1a_3 + 4g_2a_1a_2 + 3g_3a_1^3 &= 0 \qquad ; \textit{ etc.} \end{aligned}$$

The solution to these equations is

$$a_1 = \frac{1}{g_1}$$
$$a_2 = -\frac{2g_2}{g_1}a_1^2 = -\frac{2g_2}{g_1^3}$$
$$a_3 = -\frac{4g_2}{g_1^2}\left(\frac{-2g_2}{g_1^3}\right) - \frac{3g_3}{g_1^4} = \frac{8g_2^2}{g_1^5} - \frac{3g_3}{g_1^4} \qquad ; \textit{etc.}$$

Substitution into the expression for the pressure then gives

$$\frac{P}{k_B T} = g_1\left(\frac{1}{g_1 v} - \frac{2g_2}{g_1^3 v^2} + \frac{8g_2^2 - 3g_3 g_1}{g_1^5 v^3}\right) + g_2\left(\frac{1}{g_1^2 v^2} - \frac{4g_2}{g_1^4 v^3}\right)$$
$$+g_3\left(\frac{1}{g_1^3 v^3}\right) + \cdots$$

The second and third virial coefficient are then identified as

$$B(T) = \frac{g_2}{g_1^2} - \frac{2g_2}{g_1^2} = -\frac{g_2}{g_1^2}$$
$$C(T) = \frac{8g_2^2 - 3g_3 g_1}{g_1^4} - \frac{4g_2^2}{g_1^4} + \frac{g_3}{g_1^3} = \frac{4g_2^2 - 2g_3 g_1}{g_1^4}$$

This reproduces Eq. (5.133) for $B(T)$ and the stated answer for $C(T)$.

(b) The coefficients $g_l(T)$ are related to the cluster integrals B_l in Eqs. (5.104) through Eq. (5.110)

$$g_l(T) \equiv \left(\frac{2\pi m k_B T}{h^2}\right)^{3l/2}\frac{B_l}{l!}$$

Substitution into the result in part (a) for $C(T)$ then gives, with $B_1 = 1$,

$$C(T) = -\frac{2B_3}{3!} + 4\left(\frac{B_2}{2!}\right)^2 = B_2^2 - \frac{1}{3}B_3$$

It follows from Eqs. (5.101)–(5.103) that this is

$$C(T) = B_2^2 - \frac{1}{3}B_3 = -\frac{1}{3V}\int\int\int d\tau_1 d\tau_2 d\tau_3\, f_{12}f_{23}f_{31}$$

which is the desired result.[13]

[13]Since $\int d\tau_k f_{jk} = \int d\tau_k f_{lk}$ is independent of the position of the first particle, one has

$$\frac{1}{V}\int d\tau_i\left(\int d\tau_j f_{ij}\right)\left(\int d\tau_k f_{jk}\right) = \frac{1}{V^2}\int d\tau_i d\tau_l\left(\int d\tau_j f_{ij}\right)\left(\int d\tau_k f_{lk}\right)$$

Problem 5.18 Show the third virial coefficient for the hard-sphere gas in Prob. 5.15 is[14]

$$C(T) = \frac{5\pi^2\sigma^6}{18} \qquad ; \text{ hard sphere}$$

Solution to Problem 5.18

We are asked to evaluate the third virial coefficient for a hard-sphere gas

$$C(T) = -\frac{1}{3V}\int\int\int d\tau_1 d\tau_2 d\tau_3\, f_{12}f_{23}f_{13}$$

where the pair interaction functions $f_{ij} = f_{ji}$ are given by

$$\begin{aligned} f_{ij} &= e^{-U(r_{ij})/k_B T} - 1 \\ &= -1 && ;\ r_{ij} < \sigma \\ &= 0 && ;\ r_{ij} > \sigma \end{aligned}$$

Let us first do the $\int d^3r_3$ for a fixed value of $\mathbf{r}_{12}$. We only get a non-zero contribution from the region $\Gamma(r_{12})$ where (see Fig. 5.4)

$$|\mathbf{r}_{23}| < \sigma \qquad ; \ |\mathbf{r}_{13}| < \sigma$$

We thus have two segments of a sphere, each with volume (see Fig. 5.5)[15]

$$V_{\text{seg}} = \frac{\pi h}{6}(3r^2 + h^2)$$

The quantities (r, h) are identified from Fig. 5.4 as

$$\begin{aligned} h &= \sigma - \frac{1}{2}r_{12} \\ r^2 &= \sigma^2 - \frac{1}{4}r_{12}^2 \end{aligned}$$

---

[14] The integrals here are more challenging, but do-able, and fun! (Use some solid geometry.)

[15] Either look this up, or do the integration yourself.

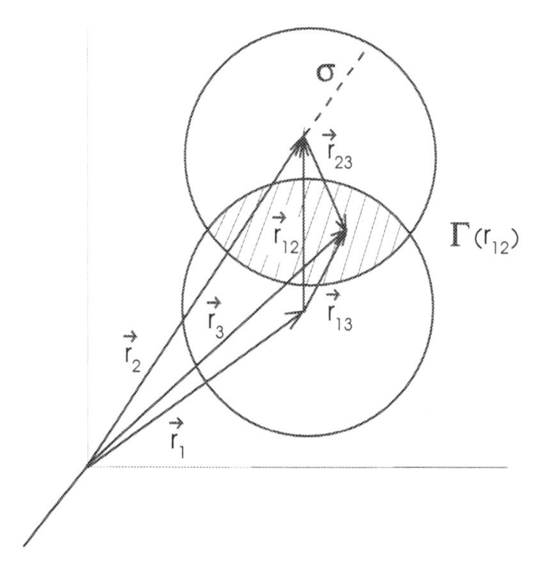

Fig. 5.4 Allowed region $\Gamma(r_{12}) = \int d^3r_3\, f_{23}f_{13}$ satisfying $|\mathbf{r}_{23}| < \sigma$ and $|\mathbf{r}_{13}| < \sigma$. In the end, we also want $|\mathbf{r}_{12}| < \sigma$.

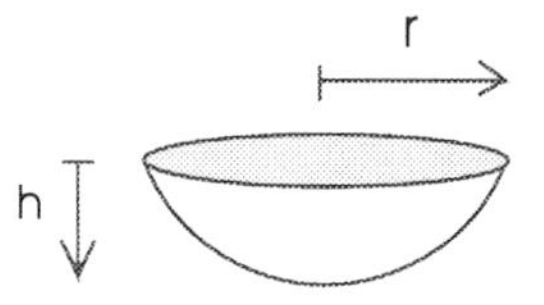

Fig. 5.5 Volume of the segment of a sphere $V_{\text{seg}} = \pi h(3r^2 + h^2)/6$.

Evaluation of $\Gamma(r_{12}) = \int d^3r_3\, f_{23}f_{13}$ then gives

$$\begin{aligned}
\Gamma(r_{12}) &= \frac{2\pi}{6}\left(\sigma - \frac{1}{2}r_{12}\right)\left(3\sigma^2 - \frac{3}{4}r_{12}^2 + \sigma^2 - \sigma r_{12} + \frac{1}{4}r_{12}^2\right) \\
&= \frac{\pi}{3}\left(\sigma - \frac{1}{2}r_{12}\right)\left(4\sigma^2 - \sigma r_{12} - \frac{1}{2}r_{12}^2\right) \\
&= \frac{\pi}{3}\left(4\sigma^3 - \sigma^2 r_{12} - \frac{1}{2}\sigma r_{12}^2 - 2\sigma^2 r_{12} + \frac{1}{2}\sigma r_{12}^2 + \frac{1}{4}r_{12}^3\right) \\
&= \frac{\pi}{3}\left(4\sigma^3 - 3\sigma^2 r_{12} + \frac{1}{4}r_{12}^3\right)
\end{aligned}$$

Thus

$$\Gamma(r_{12}) = \frac{\pi}{12}\left(16\sigma^3 - 12\sigma^2 r_{12} + r_{12}^3\right)$$

We now have to do $\int d^3r_2\, f_{12}\,\Gamma(r_{12})$ for fixed $\mathbf{r}_1$

$$\begin{aligned}\int d^3r_2\, f_{12}\,\Gamma(r_{12}) &= \int d^3r_{12} f_{12}\,\Gamma(r_{12}) \qquad ; \; r_{12} < \sigma \\ &= -4\pi \int_0^\sigma r_{12}^2 dr_{12}\,\Gamma(r_{12})\end{aligned}$$

Call $x \equiv r_{12}/\sigma$, then

$$\int d^3r_2\, f_{12} \int d^3r_3\, f_{23} f_{13} = -\frac{4\pi^2\sigma^6}{12} \int_0^1 x^2 dx\,(16 - 12x + x^3)$$

The integral is

$$\begin{aligned}\int_0^1 x^2 dx\,(16 - 12x + x^3) &= \frac{16}{3} - \frac{12}{4} + \frac{1}{6} = \frac{1}{24}(128 - 72 + 4) \\ &= \frac{5}{2}\end{aligned}$$

Hence

$$\int d^3r_2\, f_{12} \int d^3r_3\, f_{23} f_{13} = -\frac{5\pi^2\sigma^6}{6}$$

The final integral $\int d^3r_1$ just gives the volume

$$\int d^3r_1 = V$$

The third virial coefficient for the hard-sphere gas is therefore

$$\begin{aligned}C(T) &= -\frac{1}{3V} V \left(-\frac{5\pi^2\sigma^6}{6}\right) \\ &= \frac{5\pi^2\sigma^6}{18}\end{aligned}$$

which is the stated answer.

Problem 5.19 Verify the reduced form of Van der Waal's equation of state in Eq. (5.143).

Solution to Problem 5.19

Van der Waal's equation of state is given in Eq. (5.136)

$$\left(P + \frac{a}{v^2}\right)(v - b) = k_{\rm B}T$$

The values of (v, P, T) at the critical point are those in Eqs. (5.139)

$$v_c = 3b \qquad ; \; k_{\rm B}T_c = \frac{8a}{27b} \qquad ; \; P_c = \frac{a}{27b^2}$$

The reduced values of (v, P, T) are defined by the ratios in Eqs. (5.142)

$$v_r \equiv \frac{v}{v_c} \qquad ; \; P_r \equiv \frac{P}{P_c} \qquad ; \; T_r \equiv \frac{T}{T_c}$$

Divide the equation of state by $a/9b$, and factor $3b$ from $(v - b)$

$$\frac{27b^2}{a}\left(P + \frac{a}{v^2}\right)\left(v_r - \frac{1}{3}\right) = \frac{8}{3}T_r$$

This, in turn, is re-written as

$$\left(P_r + \frac{3}{v_r^2}\right)\left(v_r - \frac{1}{3}\right) = \frac{8}{3}T_r$$

Furthermore, the ratio of critical values is immediately calculated to be the pure number $3/8$

$$\frac{P_c v_c}{k_{\rm B}T_c} = \frac{3}{8}$$

These are Eqs. (5.143).[16]

[16]Compare Eq. (5.165) and Table 5.1 in the text.

Chapter 6

The Grand Canonical Ensemble

Problem 6.1 Use the first and second laws of thermodynamics to show that the thermodynamic potential satisfies the conditions in Eqs. (6.21) and (6.22).

Solution to Problem 6.1

The thermodynamic potential is given in Eqs. (6.18) and (1.18)

$$\begin{aligned}\Phi &= A - N\mu \\ &= E - TS - N\mu\end{aligned}$$

The variation is then given by[1]

$$\delta\Phi = \delta E - T\delta S - S\delta T - N\delta\mu - \mu\delta N$$

For an open system, the variation in energy follows from the first law [see Eqs. (1.26) and (1.31)]

$$\delta E = đQ - P\delta V + \mu\delta N \qquad ; \text{ first law}$$

Hence for an allowable transition at fixed (T, V, μ)

$$\delta\Phi\Big|_{T,V,\mu} = đQ - T\delta S \leq 0 \qquad ; \text{ second law}$$

where the last inequality follows from the second law [see Eq. (1.30)]. This is Eq. (6.21).

The condition for thermodynamic *equilibrium* at fixed (T, V, μ) is therefore

$$\delta\Phi\Big|_{T,V,\mu} \geq 0 \qquad ; \text{ equilbrium}$$

[1]Recall Prob. 1.9.

At equilibrium at given (T, V, μ), an assembly will minimize its thermodynamic potential. This is Eq. (6.22).

Problem 6.2 The pressure P of an assembly of non-localized systems can be obtained either from Eq. (6.20) or the last of Eqs. (6.28). Equate these two expressions, and show the logarithm of the grand partition function must be of the form

$$\ln(\text{G.P.F.}) = Vf(T, \lambda) \qquad ; \text{ non-localized systems}$$

Compare with the results in Eqs. (6.30) and (6.43).

Solution to Problem 6.2

In the grand canonical ensemble, the thermodynamic potential is given by Eq. (6.16)

$$\Phi(T, V, \mu) = -k_{\text{B}}T \ln(\text{G.P.F.}) \qquad ; \text{ grand canonical ensemble}$$

The thermodynamic potential is, in turn, related to the pressure by Eq. (6.15)

$$-PV \equiv \Phi(T, V, \mu) \qquad ; \text{ thermodynamic potential}$$

This provides one expression for the pressure in terms of the grand partition function [see Eq. (6.20)]

$$P(V, T, \lambda) = \frac{k_{\text{B}}T}{V} \ln(\text{G.P.F.}) \qquad ; \lambda = e^{\mu/k_{\text{B}}T}$$

Alternatively, with the variable set (V, T, λ), one has the thermodynamic relation in Eqs. (6.28)

$$P(V, T, \lambda) = k_{\text{B}}T \frac{\partial}{\partial V} \ln(\text{G.P.F.})$$

If these two expressions for the pressure are equated, one obtains

$$V \frac{\partial}{\partial V} \ln(\text{G.P.F.}) = \ln(\text{G.P.F.})$$

This implies that $\ln(\text{G.P.F.})$ is a homogeneous function of degree one in the volume V [2]

$$\ln(\text{G.P.F.}) = Vf(T, \lambda)$$

[2]Re-write the differential equation as $d\ln\{\ln(\text{G.P.F.})\} = d\ln V$; now integrate.

It is shown in Eq. (6.30) that for independent, non-localized systems

$$(\text{G.P.F.}) = \sum_{N=0}^{\infty} \frac{(\text{p.f.})^N}{N!} \lambda^N = \exp\left[\,\lambda(\text{p.f.})\,\right]$$

where from Eq. (2.130)

$$(\text{p.f.}) = V\left(\frac{2\pi m k_{\rm B} T}{h^2}\right)^{3/2} \qquad ; \text{ particle in box}$$

It follows that

$$\ln(\text{G.P.F.}) = \lambda(\text{p.f.}) = V\left[\lambda\left(\frac{2\pi m k_{\rm B} T}{h^2}\right)^{3/2}\right]$$

which exhibits the derived volume dependence.

For an imperfect gas, one has from Eq. (6.43)

$$(\text{G.P.F.}) = \exp\left(V\sum_{l=1}^{\infty} g_l \lambda^l\right) \qquad ; \text{ imperfect gas}$$

Hence, in this case

$$\ln(\text{G.P.F.}) = V\left[\sum_{l=1}^{\infty} g_l \lambda^l\right]$$

which again exhibits the derived volume dependence.

Problem 6.3 If E_i represents the energy of the state i in an assembly of N systems, the grand partition function can be written

$$(\text{G.P.F.}) = \sum_N \sum_i e^{-E_i/k_{\rm B}T} e^{N\mu/k_{\rm B}T}$$

(a) In quantum mechanics, the Trace of an operator represents the sum of the diagonal elements taken between a complete set of states $|\Psi_i\rangle$ in the appropriate Hilbert space. Show that if one uses simultaneous eigenstates of $(\hat{H}, \hat{N})$, then

$$(\text{G.P.F.}) = \text{Trace}\left\{e^{-(\hat{H}-\mu\hat{N})/k_{\rm B}T}\right\}$$

(b) Show that the Trace is invariant under a unitary transformation to any other complete set of states.

The expression in (a) forms the general definition of the grand partition function used as the starting point in [Fetter and Walecka (2003)].

Solution to Problem 6.3

(a) Since for a many-body assembly, the hamiltonian $\hat{H}$ and number operator $\hat{N}$ commute, one can find simultaneous eigenstates

$$\hat{H}|N,E_i\rangle = E_i|N,E_i\rangle \qquad ; \; \hat{N}|N,E_i\rangle = N|N,E_i\rangle$$

Here i denotes a complete set of quantum numbers for each E. Then

$$\begin{aligned}
(\text{G.P.F.}) &= \text{Trace}\left\{e^{-(\hat{H}-\mu\hat{N})/k_{\text{B}}T}\right\} \\
&= \sum_N \sum_i \langle N,E_i| \left\{e^{-(\hat{H}-\mu\hat{N})/k_{\text{B}}T}\right\} |N,E_i\rangle \\
&= \sum_N \sum_i e^{-E_i/k_{\text{B}}T} e^{N\mu/k_{\text{B}}T} \langle N,E_i|N,E_i\rangle \\
&= \sum_N \sum_i e^{-E_i/k_{\text{B}}T} e^{N\mu/k_{\text{B}}T}
\end{aligned}$$

(b) Suppose we compute the Trace with a different set of states $|n\rangle$

$$\text{Trace}\left\{e^{-(\hat{H}-\mu\hat{N})/k_{\text{B}}T}\right\} = \sum_n \langle n| \left\{e^{-(\hat{H}-\mu\hat{N})/k_{\text{B}}T}\right\} |n\rangle$$

Insert the completeness relation

$$\sum_N \sum_i |N,E_i\rangle\langle N,E_i| = \hat{1}$$

Then

$$\sum_n \langle n|\left\{e^{-(\hat{H}-\mu\hat{N})/k_{\text{B}}T}\right\}|n\rangle = \sum_N \sum_i e^{-E_i/k_{\text{B}}T} e^{N\mu/k_{\text{B}}T} \sum_n \langle N,E_i|n\rangle\langle n|N,E_i\rangle$$

The second set of states is also assumed to be complete

$$\sum_n |n\rangle\langle n| = \hat{1}$$

Therefore

$$\begin{aligned}
\sum_n \langle n| \left\{e^{-(\hat{H}-\mu\hat{N})/k_{\text{B}}T}\right\} |n\rangle &= \sum_N \sum_i e^{-E_i/k_{\text{B}}T} e^{N\mu/k_{\text{B}}T} \langle N,E_i|N,E_i\rangle \\
&= \sum_N \sum_i e^{-E_i/k_{\text{B}}T} e^{N\mu/k_{\text{B}}T}
\end{aligned}$$

Hence the Trace is invariant under a unitary transformation to any other complete set of states.

Problem 6.4 In analogy to the grand partition function (G.P.F.) defined at externally fixed (T, μ, V), one can define a $(\mathcal{G}.\mathcal{P}.\mathcal{F}.)$ at fixed (T, N, P)

$$(\mathcal{G}.\mathcal{P}.\mathcal{F}.) \equiv \sum_V \left(\sum_i e^{-E_i/k_{\rm B}T} \right) e^{-PV/k_{\rm B}T} = \sum_V ({\rm P.F.})_V e^{-PV/k_{\rm B}T}$$
$$= \sum_V \left(\sum_E \Omega(E, V, N) e^{-E/k_{\rm B}T} \right) e^{-PV/k_{\rm B}T}$$

where everything is computed for a given N.

Pick out the largest term in the sum, and make the identification

$$G(T, N, P) = -k_{\rm B}T \ln{(\mathcal{G}.\mathcal{P}.\mathcal{F}.)}$$

where $G(T, N, P)$ is the *Gibbs free energy* satisfying

$$dG = -SdT + VdP + \mu dN$$

Solution to Problem 6.4

Consider the $(\mathcal{G}.\mathcal{P}.\mathcal{F}.)$ as defined in this problem

$$(\mathcal{G}.\mathcal{P}.\mathcal{F}.) \equiv \sum_V \left(\sum_E \Omega(E, V, N) e^{-E/k_{\rm B}T} \right) e^{-PV/k_{\rm B}T}$$

This is now a function of (T, N, P). If just the largest term in the sum is retained, we have

$$(\mathcal{G}.\mathcal{P}.\mathcal{F}.) \doteq \Omega(E, V, N) e^{-E/k_{\rm B}T} e^{-PV/k_{\rm B}T}$$
$$-k_{\rm B}T \ln{(\mathcal{G}.\mathcal{P}.\mathcal{F}.)} = -k_{\rm B}T \, \ln{\Omega(E, V, N)} + E + PV$$

where these are now all thermodynamic variables. With the identification of the entropy as

$$S = k_{\rm B} \ln{\Omega(E, V, N)}$$

the above becomes

$$-k_{\rm B}T \ln{(\mathcal{G}.\mathcal{P}.\mathcal{F}.)} = -TS + E + PV$$
$$= A + PV = G \qquad ; \text{ Gibbs free energy}$$

where G is the Gibbs free energy. Hence

$$G(T, N, P) = -k_{\rm B}T \ln{(\mathcal{G}.\mathcal{P}.\mathcal{F}.)}$$

We know from Eq. (1.26) that the differential of G is

$$dG = -SdT + VdP + \mu dN$$

Problem 6.5 Specialize the results in Prob. 6.4 to the one-dimensional case through the identification $V \to L$(length), and $P \to -\tau$(tension).[3] Use Boltzmann statistics, and consider the following problem:

N monomeric units are arranged in a straight line to form a chain molecule. Each monomeric unit is assumed to be capable of being either in an α state or a β state. In the former state, the length is a and the energy is ε_a, while in the latter the length is b and energy ε_b (see Fig. 6.1).

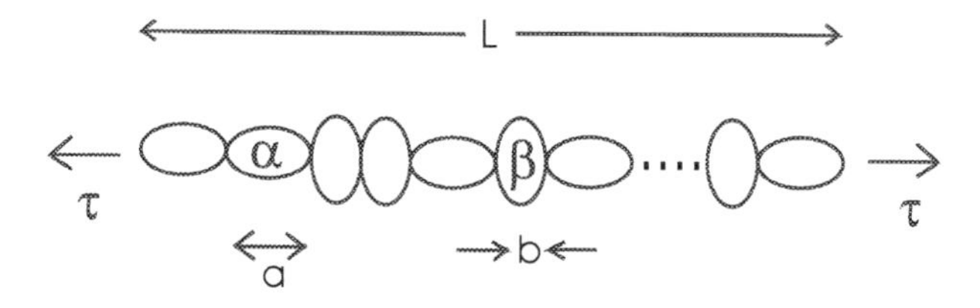

Fig. 6.1 N monomeric units arranged along a straight line to form a chain molecule. The systems have two states (α, β) with lengths (a, b) and energies $(\varepsilon_a, \varepsilon_b)$ respectively. The chain is placed under a tension τ.

(a) Compute the $(\mathcal{G.P.F.})$ by summing over (N_α, N_β). Show

$$(\mathcal{G.P.F.}) = \left[e^{-(\varepsilon_a - \tau a)/k_{\rm B}T} + e^{-(\varepsilon_b - \tau b)/k_{\rm B}T}\right]^N$$
$$G(T, N, \tau) = -Nk_{\rm B}T \ln\left[e^{-(\varepsilon_a - \tau a)/k_{\rm B}T} + e^{-(\varepsilon_b - \tau b)/k_{\rm B}T}\right]$$

(b) Show the equilibrium length of the chain is

$$\frac{L}{N} = \frac{ae^{-(\varepsilon_a - \tau a)/k_{\rm B}T} + be^{-(\varepsilon_b - \tau b)/k_{\rm B}T}}{e^{-(\varepsilon_a - \tau a)/k_{\rm B}T} + e^{-(\varepsilon_b - \tau b)/k_{\rm B}T}}$$

(c) Define the elasticity by $\chi \equiv [\partial(L/N)/\partial\tau]_{\tau=0}$. Show

$$\chi = \frac{(a-b)^2}{k_{\rm B}T} \frac{e^{-(\varepsilon_a + \varepsilon_b)/k_{\rm B}T}}{(e^{-\varepsilon_a/k_{\rm B}T} + e^{-\varepsilon_b/k_{\rm B}T})^2}$$

This is a model for the keratin molecules in wool.

[3]We use a minus sign here for the *tension*.

Solution to Problem 6.5

The results in Prob. 6.4 can be specialized to the one-dimensional case through the identification $V \to L$(length), and $P \to -\tau$(tension). Thus

$$G(T, N, \tau) = -k_{\rm B}T \ln(\mathcal{G.P.F.})$$
$$dG = -SdT - Ld\tau + \mu dN$$

where

$$(\mathcal{G.P.F.}) \equiv \sum_L \left(\sum_E \Omega(E, L, N) e^{-E/k_{\rm B}T} \right) e^{\tau L/k_{\rm B}T}$$

(a) It is clear from Fig. 6.1 that in this problem

$$N = N_\alpha + N_\beta$$
$$L = aN_\alpha + bN_\beta$$
$$E = \varepsilon_a N_\alpha + \varepsilon_b N_\beta$$

The number of ways of distributing the (N_α, N_β) types on the N sites is

$$\Omega(N_\alpha, N_\beta, N) = \frac{N!}{N_\alpha! N_\beta!} \qquad ; \; N_\alpha + N_\beta = N$$

Hence, since (N_α, N_β) now specify (E, L),

$$(\mathcal{G.P.F.}) = \sum_{N_\alpha} \sum_{N_\beta} \frac{N!}{N_\alpha! N_\beta!} e^{-(\varepsilon_a N_\alpha + \varepsilon_b N_\beta)/k_{\rm B}T} e^{\tau(aN_\alpha + bN_\beta)/k_{\rm B}T}$$

This can be re-arranged to read

$$(\mathcal{G.P.F.}) = \sum_{N_\alpha} \sum_{N_\beta} \frac{N!}{N_\alpha! N_\beta!} \left[e^{-(\varepsilon_a - a\tau)/k_{\rm B}T} \right]^{N_\alpha} \left[e^{-(\varepsilon_b - b\tau)/k_{\rm B}T} \right]^{N_\beta}$$
$$; \; N_\alpha + N_\beta = N$$

Now invoke the *binomial theorem* to obtain an exact expression for $(\mathcal{G.P.F.})$

$$(\mathcal{G.P.F.}) = \left[e^{-(\varepsilon_a - a\tau)/k_{\rm B}T} + e^{-(\varepsilon_b - b\tau)/k_{\rm B}T} \right]^N$$

It follows that

$$G(T, N, \tau) = -Nk_{\rm B}T \ln \left[e^{-(\varepsilon_a - \tau a)/k_{\rm B}T} + e^{-(\varepsilon_b - \tau b)/k_{\rm B}T} \right]$$

These are the stated answers in part (a).

(b) The equilibrium length of the chain is obtained as

$$L = -\left(\frac{\partial G}{\partial \tau}\right)_{N,T}$$

Thus

$$\frac{L}{N} = \frac{ae^{-(\varepsilon_a - \tau a)/k_{\rm B}T} + be^{-(\varepsilon_b - \tau b)/k_{\rm B}T}}{e^{-(\varepsilon_a - \tau a)/k_{\rm B}T} + e^{-(\varepsilon_b - \tau b)/k_{\rm B}T}}$$

(c) The *elasticity* is defined in the problem by

$$\chi \equiv \left[\frac{\partial(L/N)}{\partial \tau}\right]_{\tau=0}$$

This is computed to be

$$\chi = \frac{1}{k_{\rm B}T}\left\{-\left[\frac{ae^{-\varepsilon_a/k_{\rm B}T} + be^{-\varepsilon_b/k_{\rm B}T}}{e^{-\varepsilon_a/k_{\rm B}T} + e^{-\varepsilon_b/k_{\rm B}T}}\right]^2 + \frac{a^2e^{-\varepsilon_a/k_{\rm B}T} + b^2e^{-\varepsilon_b/k_{\rm B}T}}{e^{-\varepsilon_a/k_{\rm B}T} + e^{-\varepsilon_b/k_{\rm B}T}}\right\}$$

A little algebra then gives

$$\chi = \frac{(a-b)^2}{k_{\rm B}T}\,\frac{e^{-(\varepsilon_a+\varepsilon_b)/k_{\rm B}T}}{(e^{-\varepsilon_a/k_{\rm B}T} + e^{-\varepsilon_b/k_{\rm B}T})^2}$$

As stated in the problem, this is a model for the keratin molecules in wool.

Chapter 7

Applications of the Grand Canonical Ensemble

Problem 7.1 (a) Show that with the periodic boundary conditions of Eqs. (5.24)–(5.25), the normal-mode solutions of Eq. (5.23) satisfy

$$\frac{1}{V}\int_{\text{box}} e^{i(\mathbf{k}-\mathbf{k}')\cdot\mathbf{x}}\, d^3x = \delta_{\mathbf{k},\mathbf{k}'}$$

where $\delta_{\mathbf{k},\mathbf{k}'}$ is a Kronecker delta;

(b) Substitute Eqs. (7.22)–(7.23) into Eq. (7.25), do the integrals of the plane waves over the box, use the orthonormality of the unit vectors, and obtain the normal-mode expansion for electromagnetic radiation in a cavity in Eq. (7.26).[1]

Solution to Problem 7.1

(a) Since both the exponential and integration volume factor, we can focus on the one-dimensional case. Use $k_n = 2\pi n/L$ with $n = 0, \pm 1, \pm 2, \cdots$ for p.b.c., and then

$$\begin{aligned}
\frac{1}{L}\int_0^L e^{i(k-k')x}\, dx &= \frac{1}{i(k-k')L}\left[e^{i(k-k')x}\right]_0^L \\
&= \frac{1}{2\pi i(n-n')}\left[e^{2\pi i(n-n')} - 1\right] \\
&= \delta_{n,n'}
\end{aligned}$$

Hence, in three dimensions,

$$\frac{1}{V}\int_{\text{box}} e^{i(\mathbf{k}-\mathbf{k}')\cdot\mathbf{x}}\, d^3x = \delta_{\mathbf{k},\mathbf{k}'}$$

[1]See, for example, [Walecka (2008)].

(b) In c.g.s. units, in the Coulomb gauge, the electromagnetic radiation field is given in terms of the vector potential by Eqs. (7.22)–(7.23)[2]

$$\boldsymbol{\mathcal{E}} = -\frac{1}{c}\frac{\partial \mathbf{A}}{\partial t}$$
$$\mathbf{B} = \boldsymbol{\nabla} \times \mathbf{A}$$
$$\mathbf{A}(\mathbf{x}, t) = \sum_{\mathbf{k}} \sum_{s=1}^{2} \left(\frac{2\pi\hbar c^2}{\omega_k V}\right)^{1/2} \left[a_{\mathbf{k}s}\mathbf{e}_{\mathbf{k}s}e^{i(\mathbf{k}\cdot\mathbf{x}-\omega_k t)} + a_{\mathbf{k}s}^{\star}\mathbf{e}_{\mathbf{k}s}e^{-i(\mathbf{k}\cdot\mathbf{x}-\omega_k t)}\right]$$

The vector potential has been expanded in normal modes, and we apply periodic boundary conditions. The transverse unit vectors are shown in Fig. 7.2 in the text.

The total energy in the electromagnetic field is given by Eq. (7.25)

$$E = \frac{1}{8\pi}\int_{\text{box}} (\boldsymbol{\mathcal{E}}^2 + \mathbf{B}^2)\, d^3x$$

With the use of the above, the fields $(\boldsymbol{\mathcal{E}}, \mathbf{B})$ are expressed in terms of the normal-mode expansion of the vector potential $\mathbf{A}(\mathbf{x}, t)$ as

$$\boldsymbol{\mathcal{E}} = i\sum_{\mathbf{k}} \sum_{s=1}^{2} \left(\frac{2\pi\hbar\omega_k}{V}\right)^{1/2} \mathbf{e}_{\mathbf{k}s}\left[a_{\mathbf{k}s}e^{i(\mathbf{k}\cdot\mathbf{x}-\omega_k t)} - a_{\mathbf{k}s}^{*}e^{-i(\mathbf{k}\cdot\mathbf{x}-\omega_k t)}\right]$$
$$\mathbf{B} = i\sum_{\mathbf{k}} \sum_{s=1}^{2} \left(\frac{2\pi\hbar c^2}{\omega_k V}\right)^{1/2} (\mathbf{k} \times \mathbf{e}_{\mathbf{k}s})\left[a_{\mathbf{k}s}e^{i(\mathbf{k}\cdot\mathbf{x}-\omega_k t)} - a_{\mathbf{k}s}^{*}e^{-i(\mathbf{k}\cdot\mathbf{x}-\omega_k t)}\right]$$

We proceed to substitute these expressions in the relation for the field energy, and use the orthonormality of the plane waves from part (a). Consider first the term containing the contribution of the electric field

$$\frac{1}{8\pi}\int_V d^3x\, \boldsymbol{\mathcal{E}}^2 = \frac{1}{4}\sum_{\mathbf{k},s}\sum_{\mathbf{k}',s'} \hbar\omega_k \left[\delta_{\mathbf{k},\mathbf{k}'}\,\delta_{ss'}(a_{\mathbf{k}s}^{*}a_{\mathbf{k}s} + a_{\mathbf{k}s}a_{\mathbf{k}s}^{*})\right.$$
$$\left. - \delta_{\mathbf{k},-\mathbf{k}'}\,(\mathbf{e}_{\mathbf{k}s}\cdot\mathbf{e}_{-\mathbf{k},s'})\left(a_{\mathbf{k}s}a_{-\mathbf{k},s'}\,e^{-2i\omega_k t} + a_{\mathbf{k}s}^{*}a_{-\mathbf{k},s'}^{*}\,e^{2i\omega_k t}\right)\right]$$

Now consider the term containing the contribution of the magnetic field

$$\frac{1}{8\pi}\int_V d^3x\, \mathbf{B}^2 = \frac{1}{4}\sum_{\mathbf{k},s}\sum_{\mathbf{k}',s'} \frac{\hbar c^2}{\omega_k} \left[\delta_{\mathbf{k},\mathbf{k}'}\,\mathbf{k}^2\,\delta_{ss'}(a_{\mathbf{k}s}^{*}a_{\mathbf{k}s} + a_{\mathbf{k}s}a_{\mathbf{k}s}^{*})\right.$$
$$\left. - \delta_{\mathbf{k},-\mathbf{k}'}\,(\mathbf{k}\times\mathbf{e}_{\mathbf{k}s})\cdot(\mathbf{k}'\times\mathbf{e}_{\mathbf{k}'s'})\left(a_{\mathbf{k}s}a_{-\mathbf{k},s'}\,e^{-2i\omega_k t} + a_{\mathbf{k}s}^{*}a_{-\mathbf{k},s'}^{*}\,e^{2i\omega_k t}\right)\right]$$

[2]Remember that in this book we work in c.g.s. units. Here we write the electric field as $\boldsymbol{\mathcal{E}}$ so as not to confuse it with the energy, and we start with the classical expression for $\mathbf{A}(\mathbf{x}, t)$, before quantization (see later).

Use the vector identity[3]

$$(\mathbf{k} \times \mathbf{e}_{\mathbf{k}s}) \cdot (\mathbf{k} \times \mathbf{e}_{-\mathbf{k},s'}) = \mathbf{k}^2(\mathbf{e}_{\mathbf{k}s} \cdot \mathbf{e}_{-\mathbf{k},s'})$$

and recall $\omega_k^2 = \mathbf{k}^2 c^2$, so that

$$\frac{\mathbf{k}^2}{\omega_k} = \frac{\omega_k}{c^2}$$

The magnetic contribution then takes the form

$$\frac{1}{8\pi}\int_V d^3x\, \mathbf{B}^2 = \frac{1}{4}\sum_{\mathbf{k},s}\sum_{\mathbf{k}',s'} \hbar\omega_k \left[\delta_{\mathbf{k},\mathbf{k}'}\delta_{ss'}(a^*_{\mathbf{k}s}a_{\mathbf{k}s} + a_{\mathbf{k}s}a^*_{\mathbf{k}s})\right.$$
$$\left. + \,\delta_{\mathbf{k},-\mathbf{k}'}\left(\mathbf{e}_{\mathbf{k}s} \cdot \mathbf{e}_{-\mathbf{k},s'}\right)\left(a_{\mathbf{k}s}a_{-\mathbf{k},s'}\, e^{-2i\omega_k t} + a^*_{\mathbf{k}s}a^*_{-\mathbf{k},s'}\, e^{2i\omega_k t}\right)\right]$$

When the electric and magnetic contributions are combined, the final time-dependent terms cancel identically, and the energy in the field is then indeed given by the normal-mode expansion

$$E = \frac{1}{2}\sum_{\mathbf{k}}\sum_{s=1}^{2} \hbar\omega_k\,(a^*_{\mathbf{k}s}a_{\mathbf{k}s} + a_{\mathbf{k}s}a^*_{\mathbf{k}s})$$

This represents an infinite set of uncoupled simple harmonic oscillators, and the oscillators are quantized in the canonical fashion. The normal-mode amplitudes $(a^*_{\mathbf{k}s}, a_{\mathbf{k}s})$ become creation and destruction operators $(a^\dagger_{\mathbf{k}s}, a_{\mathbf{k}s})$, and the energy operator for the assembly is then given by Eq. (7.26)

$$E = \frac{1}{2}\sum_{\mathbf{k}}\sum_{s=1}^{2} \hbar\omega_k \left(a^\dagger_{\mathbf{k}s}a_{\mathbf{k}s} + a_{\mathbf{k}s}a^\dagger_{\mathbf{k}s}\right)$$
$$= \sum_{\mathbf{k}}\sum_{s=1}^{2} \hbar\omega_k \left(a^\dagger_{\mathbf{k}s}a_{\mathbf{k}s} + \frac{1}{2}\right)$$

The eigenvalues of this operator give the energies of the quantized radiation field

$$E_{\text{tot}} = \sum_{\mathbf{k}}\sum_{s=1}^{2} \hbar\omega_k \left(n_{\mathbf{k}s} + \frac{1}{2}\right) \qquad ; \; n_{\mathbf{k}s} = 0, 1, 2, \cdots, \infty$$

Problem 7.2 Show that the total energy E can be written in the form in Eq. (7.20) for all three cases of Boltzmann, Bose-Einstein, and Fermi-

[3]Note that $(\mathbf{k}' \times \mathbf{e}_{\mathbf{k}'s'}) \to -(\mathbf{k} \times \mathbf{e}_{-\mathbf{k},s'})$; note also that $(\mathbf{k} \cdot \mathbf{e}_{-\mathbf{k}s'}) = 0$ for $s' = (1, 2)$.

Dirac statistics. Recall that both (V, λ) are to be kept constant in taking the partial derivative in Eq. (7.19).

Solution to Problem 7.2

The grand partition functions for Boltzmann, Bose-Einstein, and Fermi-Dirac statistics are given in Eqs. (7.6), (7.13), and (7.15) respectively[4]

$$
\begin{aligned}
(\text{G.P.F.})_{\text{Boltz}} &= \prod_i \exp\left[\lambda e^{-\varepsilon_i/k_{\text{B}}T}\right] \\
(\text{G.P.F.})_{\text{Bose}} &= \prod_i \left(\frac{1}{1-\lambda e^{-\varepsilon_i/k_{\text{B}}T}}\right) \\
(\text{G.P.F.})_{\text{Fermi}} &= \prod_i \left(1+\lambda e^{-\varepsilon_i/k_{\text{B}}T}\right)
\end{aligned}
$$

The internal energy follows from Eq. (7.19)

$$
E = k_{\text{B}}T^2 \frac{\partial}{\partial T} \ln(\text{G.P.F.})
$$

where it follows from Eqs. (6.28) that the derivative is to be carried out at fixed (V, λ). The ln of a product is the sum of the ln's, and this gives

$$
\begin{aligned}
E_{\text{Boltz}} &= \sum_i \varepsilon_i \left(\lambda e^{-\varepsilon_i/k_{\text{B}}T}\right) \\
E_{\text{Bose}} &= \sum_i \varepsilon_i \left(\frac{\lambda e^{-\varepsilon_i/k_{\text{B}}T}}{1-\lambda e^{-\varepsilon_i/k_{\text{B}}T}}\right) \\
E_{\text{Fermi}} &= \sum_i \varepsilon_i \left(\frac{\lambda e^{-\varepsilon_i/k_{\text{B}}T}}{1+\lambda e^{-\varepsilon_i/k_{\text{B}}T}}\right)
\end{aligned}
$$

The results are all of the form in Eq. (7.20)

$$
E = \sum_i \varepsilon_i n_i^\star
$$

with the $n_i^\star$ of Eqs. (7.18).

Problem 7.3 (a) Compare with the Debye theory of the heat capacity of a crystal, and explain why one is always in the T^3-regime for the heat capacity of a photon gas;

(b) Explain the limiting procedure required to get from Eq. (5.35) to Eq. (7.44).

---

[4]Here $\lambda = e^{\mu/k_{\text{B}}T}$ is the absolute activity [see Eq. (6.8)].

Solution to Problem 7.3

(a) The quanta in both the phonon gas in a crystal and the photon gas obey the Bose distribution law with vanishing chemical potential and have the quadratic spectral distributions in Eqs. (5.27) and (7.37)

$$\frac{1}{V}g(\nu)d\nu = \frac{4\pi\nu^2 d\nu}{c_s^3} \qquad ; \text{ phonons}$$
$$= \frac{8\pi\nu^2 d\nu}{c^3} \qquad ; \text{ photons}$$

The only differences are the appearance of the sound velocity c_s for phonons, and our inclusion of the two transverse modes from the outset for photons.

In the photon case, there is no cutoff on the frequency, $\nu_m \to \infty$, and hence the upper limit on the frequency integral is always infinite in Eq. (5.33), and one is always in the T^3 regime with photons.

(b) In the phonon case, we scaled out the Debye temperature in Eq. (5.31)

$$\theta_{\rm D} = \frac{h\nu_m}{k_{\rm B}}$$

With photons, we no longer want to do this. We leave the spectral density in Eq. (5.29) just as above

$$9\left(\frac{N}{V}\right)\frac{\nu^2 d\nu}{\nu_m^3} = \frac{8\pi\nu^2 d\nu}{c^3}$$

This implies

$$9\left(\frac{N}{V}\right)\frac{1}{(k_{\rm B}\theta_{\rm D})^3} = \frac{8\pi}{(hc)^3} \qquad ; \nu_m \to \infty$$

Note that it is the *ratio* of N/V to ν_m^3 that remains finite as $\nu_m \to \infty$; there is no longer a finite number N of underlying sytems in a lattice.

The result in Eq. (5.35) for phonons is then re-written for photons as

$$\frac{1}{V}C_V = \frac{12\pi^4}{5}\left(\frac{N}{V}\right)k_{\rm B}\left(\frac{T}{\theta_{\rm D}}\right)^3 \qquad ; \text{ phonons}$$
$$= \left[\frac{4\pi^2 k_{\rm B}^4}{15(\hbar c)^3}\right]T^3 \qquad ; \text{ photons}$$

This now agrees with Eq. (7.44).[5]

[5] Note the presence of $\hbar = h/2\pi$ in the last result.

Problem 7.4 (a) Show the pressure $P(T)$ exerted by a photon gas is

$$P(T) = \left[\frac{\pi^2 k_B^4}{45(\hbar c)^3}\right] T^4$$

(b) Put in numbers and compute P/T^4 in dynes/cm^2 $^\circ$K^4, and also in atm/ $^\circ$K^4. What is P at 300 $^\circ$K? At 3000 $^\circ$K? At 3×10^6 $^\circ$K?

Solution to Problem 7.4

(a) In Eqs. (7.45)–(7.47) the pressure of a photon gas is computed starting from the relation

$$PV = k_B T \ln (\text{G.P.F.})$$

With the use of the appropriate spectral density, the Bose distribution at vanishing chemical potential, and a partial integration, it is demonstrated that the photon gas satisfies the equation of state

$$PV = \frac{1}{3}E$$

The energy density of the photon gas is obtained in Eq. (7.43)

$$\begin{aligned}
\frac{E}{V} &= \frac{(k_B T)^4}{\pi^2(\hbar c)^3}\int_0^\infty \frac{x^3\,dx}{e^x - 1} \\
&= \left[\frac{\pi^2 k_B^4}{15(\hbar c)^3}\right] T^4 \qquad ;\ \text{photon energy}
\end{aligned}$$

The pressure follows as

$$P(T) = \left[\frac{\pi^2 k_B^4}{45(\hbar c)^3}\right] T^4 \qquad ;\ \text{photon pressure}$$

(b) The energy density of the photon gas is expressed in terms of the Stefan-Boltzmann constant σ according to[6]

$$\begin{aligned}
\frac{E}{V} &= \frac{4\sigma}{c}T^4 \\
\sigma &= \frac{\pi^2 k_B^4}{60\hbar^3 c^2} = 5.670 \times 10^{-5}\,\frac{\text{erg}}{\text{sec-cm}^2}\,\frac{1}{^\circ\text{K}^4}
\end{aligned}$$

Then, with $c = 2.998 \times 10^{10}$ cm/sec, the pressure becomes

$$\frac{P(T)}{T^4} = \frac{4\sigma}{3c} = 2.522 \times 10^{-15}\,\frac{\text{dyne}}{\text{cm}^2}\,\frac{1}{^\circ\text{K}^4}$$

[6] See, for example, [Walecka (2008)].

The pressure of the photon gas at various temperatures is shown in Table 7.1 in both dynes/cm^2 and atmospheres (atm). Note that while the radiation pressure is tiny at room temperature, at a temperature of $3 \times 10^6\,^{\circ}\mathrm{K}$ it is comparable to the highest pressure achieved in the laboratory.

Table 7.1 Pressure P of a photon gas at various temperatures T. Note that $1\,\mathrm{atm} = 1.013 \times 10^6\,\mathrm{dynes/cm}^2$.

$T(^{\circ}K)$	$P(\mathrm{dynes/cm}^2)$	$P(\mathrm{atm})$
300	2.043×10^{-5}	2.017×10^{-11}
3000	2.043×10^{-1}	2.017×10^{-7}
3×10^6	2.043×10^{11}	2.017×10^{5}

Problem 7.5 Use Fig. 5.6 in the text to explain the exponential fall-off in Eq. (7.42).

Solution to Problem 7.5

When the thermal energy is much greater than that of a photon $k_{\mathrm{B}}T \gg h\nu$, there is plenty of thermal energy to excite the radiation, and the equipartition theorem states that in this limit one will have a thermal energy of $k_{\mathrm{B}}T$ per mode. Hence the radiation energy density in a cavity takes the form in Eq. (7.41)

$$u(\nu, T) = \frac{8\pi\nu^2}{c^3} k_{\mathrm{B}}T \qquad ; \frac{h\nu}{k_{\mathrm{B}}T} \ll 1$$

On the other hand, if the thermal energy is much less than that of a photon $k_{\mathrm{B}}T \ll h\nu$ as sketched in Fig. 5.6 in the text, there is insufficient thermal energy to excite the radiation, and the Bose (Planck) distribution implies the exponential fall-off in the energy density given in Eq. (7.42)

$$u(\nu, T) = \frac{8\pi\nu^2}{c^3} h\nu\, e^{-h\nu/k_{\mathrm{B}}T} \qquad ; \frac{h\nu}{k_{\mathrm{B}}T} \gg 1$$

Problem 7.6 Show that there is no Bose-Einstein condensation at any finite temperture for a two-dimensional ideal Bose gas.[7]

[7]Compare Probs. 7.22–7.23.

Solution to Problem 7.6

The particle density for an ideal Bose gas in three-dimensions is given in Eqs. (7.60)

$$\frac{N}{V} = \frac{g_s}{4\pi^2}\left(\frac{2m}{\hbar^2}\right)^{3/2}\int_0^\infty \frac{\varepsilon^{1/2}\,d\varepsilon}{e^{(\varepsilon-\mu)/k_{\rm B}T}-1} \qquad ; \text{ three dimensions}$$

Bose-Einstein condensation in this gas is discussed in section 7.3.2.2. One looks for the temperature T_0 at which the chemical potential first reaches zero, as illustrated in Fig. 7.6 in the text

$$\frac{N}{V} = \frac{g_s}{4\pi^2}\left(\frac{2m}{\hbar^2}\right)^{3/2}\int_0^\infty \frac{\varepsilon^{1/2}\,d\varepsilon}{e^{\varepsilon/k_{\rm B}T_0}-1} \qquad ; \text{ determines } T_0$$

With the introduction of the dimensionless variable $x = \varepsilon/k_{\rm B}T_0$, this becomes

$$\frac{N}{V} = \frac{g_s}{4\pi^2}\left(\frac{2mk_{\rm B}T_0}{\hbar^2}\right)^{3/2}\int_0^\infty \frac{x^{1/2}\,dx}{e^{x}-1}$$

Below this temperature, one obtains less then the actual particle density with vanishing chemical potential, and the interpretation is that there is Bose condensation below T_0, with a finite fraction of the particles going into the ground state. The dimensionless definite integral over x in the above is well-defined and convergent, since for small x the integrand behaves as $dx/\sqrt{x}$; it is evaluated in Eq. (7.68).

The particle density for the Bose gas in two dimensions follows as in Prob. 7.23(c)

$$\frac{N}{\mathcal{A}} = \frac{g_s}{4\pi}\left(\frac{2m}{\hbar^2}\right)\int_0^\infty \frac{d\varepsilon}{e^{(\varepsilon-\mu)/k_{\rm B}T}-1} \qquad ; \text{ two dimensions}$$

An argument analogous to the above would, upon setting $\mu = 0$, lead to

$$\frac{N}{\mathcal{A}} = \frac{g_s}{4\pi}\left(\frac{2mk_{\rm B}T_0}{\hbar^2}\right)\int_0^\infty \frac{dx}{e^{x}-1} \qquad ; \underline{\text{not}} \text{ valid}$$

This relation is <u>not</u> valid. The integral is not convergent since the integrand, because of the different spectrum in two dimensions, behaves as dx/x as $x \to 0$. As a consequence, one must leave the term $e^{-\mu/k_{\rm B}T}$ in the denominator of the expression for $N/\mathcal{A}$. This protects the integrand at the origin, and with a small, negative, non-zero μ, one can match any given value of $N/\mathcal{A}$. Hence the chemical potential never reaches zero as in Fig. 7.6 in the text, and *there is no Bose-Einstein condensation at any finite temperature for a two-dimensional ideal Bose gas.*

Problem 7.7 (a) Start from Eqs. (7.60), and explain why the curve in Fig. 7.7 in the text goes to 3/2 at high T;

(b) Repeat for Eqs. (7.108) and Fig. 7.11 in the text.

Solution to Problem 7.7

(a) At high temperature, the chemical potential of the non-relativistic Bose gas of N systems behaves as $\mu/k_{\rm B}T \to -\infty$ (see Fig. 7.6 in the text). The energy and number of systems in Eqs. (7.60) then go over to the classical Boltzmann expressions

$$\frac{E}{V} \to \frac{g_s}{4\pi^2}\left(\frac{2m}{\hbar^2}\right)^{3/2} e^{\mu/k_{\rm B}T}\int_0^\infty \varepsilon^{3/2}e^{-\varepsilon/k_{\rm B}T}\,d\varepsilon \qquad ; T\to\infty$$

$$\frac{N}{V} \to \frac{g_s}{4\pi^2}\left(\frac{2m}{\hbar^2}\right)^{3/2} e^{\mu/k_{\rm B}T}\int_0^\infty \varepsilon^{1/2}e^{-\varepsilon/k_{\rm B}T}\,d\varepsilon$$

The ratio becomes

$$\frac{E}{N} = \frac{\int_0^\infty x^{3/2}e^{-x}\,dx}{\int_0^\infty x^{1/2}e^{-x}\,dx}\,k_{\rm B}T$$

Introduce the Γ-function from Eq. (2.176)

$$\Gamma(z) = \int_0^\infty e^{-t}\,t^{z-1}dt$$

The ratio E/N then takes the form

$$\frac{E}{N} = \frac{\Gamma(5/2)}{\Gamma(3/2)}\,k_{\rm B}T = \frac{3}{2}k_{\rm B}T$$

The equation of state of this Bose gas in Eq. (7.58) then becomes

$$PV = \frac{2}{3}E = Nk_{\rm B}T$$

This is the perfect gas law, and we know from Prob. 2.7 that the molar specific heat of the perfect gas is

$$C_V = \frac{3}{2}R \qquad ; T\to\infty$$

(b) The high-temperature limit of the chemical potential of the non-relativistic Fermi gas has exactly the same behavior as that of the Bose gas in part (a) (see Fig. 7.8 in the text). Thus exactly the same analysis applies to $(E/V, N/V)$ in Eqs. (7.108).[8] Hence, one again obtains the molar specific heat of the perfect gas

$$C_V = \frac{3}{2}R \qquad ; T \to \infty$$

Problem 7.8 (a) Use the density in Eq. (7.97) appropriate to the spin-zero boson ^{4}He with mass $m_{\rm He} = 6.64 \times 10^{-24}\,$gm. Invert the last of Eqs. (7.60) numerically to find $\mu(N/V, T)$;

(b) Use the results from part (a), and make a good plot of the two curves in Fig. 7.6 in the text.

Solution to Problem 7.8

(a) The chemical potential for a non-interacting Bose gas is obtained by the inversion of the last of Eqs. (7.60)

$$n = \frac{g_s}{4\pi^2}\left(\frac{2m}{\hbar^2}\right)^{3/2} \int_0^\infty \frac{\varepsilon^{1/2}\, d\varepsilon}{e^{(\varepsilon-\mu)/k_{\rm B}T} - 1}$$

The introduction of the dimensionless integration variable $x \equiv \varepsilon/k_{\rm B}T$ gives

$$\frac{n}{(k_{\rm B}T)^{3/2}} = \frac{g_s}{4\pi^2}\left(\frac{2m}{\hbar^2}\right)^{3/2} \int_0^\infty \frac{x^{1/2}\, dx}{e^{(x-\mu/k_{\rm B}T)} - 1}$$

The transition temperature T_0 is the place where the chemical potential first vanishes [see Eqs. (7.71) and (7.97)]

$$\frac{n}{(k_{\rm B}T_0)^{3/2}} = \frac{g_s}{4\pi^2}\left(\frac{2m}{\hbar^2}\right)^{3/2} \int_0^\infty \frac{x^{1/2}\, dx}{e^x - 1}$$

Take the *ratio* of these two relations, and use[9]

$$\int_0^\infty \frac{x^{1/2}\, dx}{e^x - 1} = \Gamma\left(\frac{3}{2}\right)\zeta\left(\frac{3}{2}\right) = \frac{\sqrt{\pi}}{2} 2.612$$

This gives

$$\left(\frac{T_0}{T}\right)^{3/2} = \frac{1}{\Gamma(3/2)\zeta(3/2)} \int_0^\infty \frac{x^{1/2}\, dx}{e^{(x-\mu/k_{\rm B}T)} - 1}$$

[8]The additional ∓ 1 in the denominators of the distribution functions become irrelevant in this limit.

[9]See Eqs. (7.68)–(7.69).

Note that in this form, all the constants in the problem enter only through T_0. The numerical inversion of this relation obtained with Mathcad 7 to give μ/k_BT as a function of T/T_0 is shown in Fig. 7.1. The numerical result is extrapolated down to $T/T_0 = 1$ using Eq. (7.89)

$$\mu = -k_BT_0\left[\frac{\Gamma(3/2)\zeta(3/2)}{\pi}\right]^2\left[\left(\frac{T}{T_0}\right)^{3/2} - 1\right]^2 \qquad ; \frac{T}{T_0} \to 1^+$$

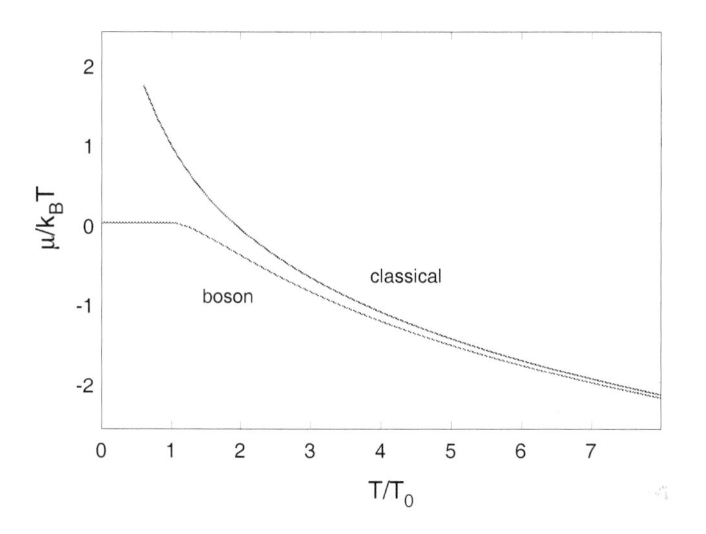

Fig. 7.1 The chemical potential μ/k_BT for non-interacting bosons obtained by numerical inversion of the relation $(T_0/T)^{3/2} = [\Gamma(3/2)\zeta(3/2)]^{-1}\int_0^\infty x^{1/2}\,dx/[e^{(x-\mu/k_BT)} - 1]$. Here T_0 is the transition temperature where $\mu = 0$. The classical limit is obtained for $\mu/k_BT \to -\infty$ [see part (b)].

(b) As demonstrated in Prob. 7.7, the classical limit is obtained for $\mu/k_BT \to -\infty$. From the above, this gives[10]

$$\left(\frac{T_0}{T}\right)^{3/2} \to \frac{1}{\Gamma(3/2)\zeta(3/2)}e^{\mu_{\rm cl}/k_BT}\int_0^\infty x^{1/2}\,e^{-x}\,dx$$
$$= \frac{1}{\zeta(3/2)}e^{\mu_{\rm cl}/k_BT}$$

[10] Use $\int_0^\infty x^{1/2}e^{-x}\,dx = \Gamma(3/2) = \sqrt{\pi}/2$.

It follows that

$$\frac{\mu_{\rm cl}}{k_{\rm B}T} = \ln\left[\zeta\left(\frac{3}{2}\right)\left(\frac{T_0}{T}\right)^{3/2}\right]$$

This curve is also shown in Fig. 7.1, which should be compared with Fig. 7.6 in the text. For liquid ^{4}He, the calculated T_0 is given in Eq. (7.97).

Problem 7.9 Define the mean value of k^n in the ground state of a Fermi gas by $\langle k^n\rangle_{\rm FG} \equiv \int_0^{k_{\rm F}} k^n d^3k / \int_0^{k_{\rm F}} d^3k$. Compute this mean value for $n = 2, 4, 6$.

Solution to Problem 7.9

Do the integral in the numerator

$$I_n \equiv \int_0^{k_{\rm F}} 4\pi k^{n+2} dk = \frac{4\pi}{n+3} k_{\rm F}^{n+3}$$

The desired mean values are then given in terms of I_n as follows:

$$\langle k^2\rangle_{\rm FG} = \frac{I_2}{I_0} = \frac{3}{5}k_{\rm F}^2$$
$$\langle k^4\rangle_{\rm FG} = \frac{I_4}{I_0} = \frac{3}{7}k_{\rm F}^4$$
$$\langle k^6\rangle_{\rm FG} = \frac{I_6}{I_0} = \frac{3}{9}k_{\rm F}^6$$

Problem 7.10 Consider a non-interacting, non-relativistic Fermi gas of spin-1/2 particles in its ground state at a given particle density $n = N/V$ (electrons in a metal, neutrons in a neutron star, *etc.*). The particles have mass m and magnetic moment μ_0. A uniform magnetic field $\mathbf{B}$ is applied. Treat the assembly as two separate Fermi gases, one with magnetic moment aligned with the field, and one with magnetic moment opposed. Parameterize the number of particles of each type as

$$N_\uparrow = \frac{N}{2}(1+\delta) \qquad ; \; N_\downarrow = \frac{N}{2}(1-\delta)$$

(a) What is the contribution to the energy of the assembly coming from the interaction of the spin magnetic moments with the magnetic field for a given δ?

(b) The kinetic energy of the assembly must increase to achieve a configuration with finite δ. Let $\varepsilon_{\rm F}^0$ be the value of the Fermi energy with $\delta = 0$.

Show the increase in kinetic energy is given to order δ^2 by[11]

$$\Delta E = \frac{\delta^2}{3}\varepsilon_{\rm F}^0 N$$

(c) Construct the total change in the energy as a sum of the two contributions in (a) and (b). Minimize with respect to δ to find the new ground state. Show

$$\delta = \frac{3\mu_0 B}{2\varepsilon_{\rm F}^0}$$

(d) The magnetic spin susceptibility is defined in terms of the magnetic dipole moment per unit volume $\mathbf{M}$ according to $\mathbf{M} = \kappa_m \mathbf{B}$. Hence re-derive the expression for the *Pauli paramagnetic spin susceptibility at zero temperature*

$$\kappa_{\rm Pauli}(0) = \frac{3\mu_0^2}{2\varepsilon_{\rm F}^0}\left(\frac{N}{V}\right)$$

Solution to Problem 7.10[12]

(a) The interaction energy of an intrinsic magnetic moment with a magnetic field $\mathbf{B}$ is $H' = -\boldsymbol{\mu}\cdot\mathbf{B}$. The total interaction energy of the system of spin-1/2 particles is therefore

$$\begin{aligned}\Delta H' &= -\mu_0 B(N_\uparrow - N_\downarrow) \\ &= -\mu_0 B\frac{N(1+\delta)}{2} + \mu_0 B\frac{N(1-\delta)}{2} = -\mu_0 BN\delta\end{aligned}$$

where we have fixed the z-direction along $\mathbf{B}$.

(b) Equation (7.116) relates the number of particles to the maximum wavenumber k_F. In line with the suggestion, we treat the system as made up of two separate Fermi gases and write[13]

$$\begin{aligned}N_\uparrow &= \frac{N(1+\delta)}{2} = \frac{V}{(2\pi)^3}\frac{4\pi}{3}k_{F\uparrow}^3 \\ N_\downarrow &= \frac{N(1-\delta)}{2} = \frac{V}{(2\pi)^3}\frac{4\pi}{3}k_{F\downarrow}^3\end{aligned}$$

---

[11] Use the Taylor series expansion $(1+x)^n = 1+nx+n(n-1)x^2/2!+\cdots$, which holds for $|x|<1$ and any n (integer or non-integer).

[12] The solutions to Probs. 7.10 and 7.12 are taken from [Amore and Walecka (2013)].

[13] Each Fermi gas now has $g_s = 1$.

These equations can be inverted to give

$$k_{F\uparrow} = \left[6\pi^2 \frac{N}{V}\frac{(1+\delta)}{2}\right]^{1/3}$$

$$k_{F\downarrow} = \left[6\pi^2 \frac{N}{V}\frac{(1-\delta)}{2}\right]^{1/3}$$

The total kinetic energy of the system will then be [see Eq. (7.119)]

$$\begin{aligned} E = E_\uparrow + E_\downarrow &= \frac{3}{5}\frac{N(1+\delta)}{2}\frac{\hbar^2 k_{F\uparrow}^2}{2m} + \frac{3}{5}\frac{N(1-\delta)}{2}\frac{\hbar^2 k_{F\downarrow}^2}{2m} \\ &= \frac{3}{10}\frac{N\hbar^2}{2m}\left(3\pi^2\frac{N}{V}\right)^{2/3}\left[(1+\delta)^{5/3} + (1-\delta)^{5/3}\right] \end{aligned}$$

For $|\delta| \ll 1$ we may expand the factors $(1\pm\delta)^{5/3}$ up to order δ^2, as explained in the footnote, to obtain

$$\begin{aligned} E &\approx \frac{3}{10}\frac{N\hbar^2}{2m}\left(3\pi^2\frac{N}{V}\right)^{2/3}\left[2 + \frac{10}{9}\delta^2\right] \\ &= \frac{3}{5}\varepsilon_\mathrm{F}^0 N\left(1 + \frac{5}{9}\delta^2\right) \end{aligned}$$

and therefore

$$\Delta E = \frac{\delta^2}{3}\varepsilon_\mathrm{F}^0 N$$

(c) The total change in energy is

$$\Delta E_\mathrm{tot} = \Delta H' + \Delta E = -\mu_0 B N\delta + \frac{\delta^2}{3}\varepsilon_\mathrm{F}^0 N$$

which describes a parabola in δ, with a minimum located at[14]

$$\delta = \frac{3\mu_0 B}{2\varepsilon_\mathrm{F}^0} \qquad ; \text{ for minimum}$$

If we use this value in ΔE_tot we obtain

$$\Delta E_\mathrm{tot} = -\frac{3}{4}\frac{N(\mu_0 B)^2}{\varepsilon_\mathrm{F}^0}$$

Since $\Delta E_\mathrm{tot} < 0$, we conclude that in a constant magnetic field, the new ground state of a system of N spin-1/2 particles with mass m and intrinsic

[14]Clearly, this result is valid only for $|\delta| \ll 1$, that is, for $|\mu_0 B|/\varepsilon_\mathrm{F}^0 \ll 1$.

magnetic moment μ_0 will have

$$N_{\uparrow} \approx \left(1 + \frac{3\mu_0 B}{2\varepsilon_{\rm F}^0}\right)\frac{N}{2} \qquad ; \; N_{\downarrow} \approx \left(1 - \frac{3\mu_0 B}{2\varepsilon_{\rm F}^0}\right)\frac{N}{2}$$

(d) The magnetic moment per unit volume of the system is

$$M = \frac{\mu_0 N\delta}{V} = \frac{3\mu_0^2 nB}{2\varepsilon_{\rm F}^0}$$

Thus the Pauli paramagnetic spin susceptibility is given by

$$\chi_{\rm Pauli}(0) = \frac{3\mu_0^2}{2\varepsilon_{\rm F}^0} n$$

which is the anticipated result.

Problem 7.11 Nuclear matter is a hypothetical material of uniform density filling a big box with periodic boundary conditions. It consists of four types of nucleons ($n\uparrow, n\downarrow, p\uparrow, p\downarrow$), with a baryon number of $B \equiv N + Z$ where N is the total number of neutrons and Z is the number of protons. The Coulomb interaction is turned off and nuclear matter is assumed to form a degenerate non-relativistic Fermi gas.[15]

(a) Show the baryon density of nuclear matter is related to the Fermi momentum $\hbar k_{\rm F}$ by

$$\frac{B}{V} = \frac{2k_{\rm F}^3}{3\pi^2} \qquad ; \text{ nuclear matter}$$

(b) The observed Fermi wavenumber of nuclear matter, inferred from measurements of proton densities through electron scattering, is

$$k_{\rm F} \approx 1.42\,{\rm F}^{-1} \qquad ; \; 1\,{\rm F} \equiv 10^{-13}\,{\rm cm}$$

Compute the Fermi energy $\varepsilon_{\rm F} = \hbar^2 k_{\rm F}^2/2m_p$ of nuclear matter in MeV;

(c) Compute the baryon density of nuclear matter.

Solution to Problem 7.11

(a) For a Fermi gas in its ground state with a spin-isospin degeneracy of $g_s = 4$, the number of filled states is from Eq. (7.115) [compare Fig. 7.10 in the text]

$$B = \frac{4V}{(2\pi)^3}\int_0^{k_{\rm F}} d^3k = \frac{2V}{3\pi^2}k_{\rm F}^3$$

[15] To a first approximation, it is the substance at the center of the Pb nucleus.

Hence the baryon density of nuclear matter is

$$\frac{B}{V} = \frac{2k_{\rm F}^3}{3\pi^2} \qquad ; \text{ nuclear matter}$$

(b) The Fermi energy of nuclear matter, the energy of the last filled level, is

$$\varepsilon_{\rm F} = \frac{\hbar^2 k_{\rm F}^2}{2m_p}$$

where m_p is the proton (nucleon) mass. Use[16]

$$\frac{\hbar^2}{2m_p} = 20.74\,\text{MeV-F}^2 \qquad ; \; 1\,\text{F} \equiv 10^{-13}\,\text{cm}$$

Then with a value $k_{\rm F} \approx 1.42\,\text{F}^{-1}$ inferred from the central proton density in heavy nuclei measured in electron scattering, one has for the Fermi energy

$$\varepsilon_{\rm F} \approx 41.8\,\text{MeV} \qquad ; \; k_{\rm F} \approx 1.42\,\text{F}^{-1}$$

(c) The corresponding baryon density is[17]

$$\frac{B}{V} \approx 0.193\,\text{F}^{-3}$$

Problem 7.12 Assume that the neutron and proton densities in nuclear matter are now driven apart by the Coulomb interaction of the protons. Write

$$N = \frac{B}{2}(1+\delta) \qquad ; \; Z = \frac{B}{2}(1-\delta)$$

(a) Follow the analysis in Prob. 7.10, and show that the kinetic energy of nuclear matter at fixed baryon density, to leading order in δ, is increased by an amount

$$\frac{\Delta E}{B} = \frac{\varepsilon_{\rm F0}}{3}\delta^2$$

Here $\varepsilon_{\rm F0}$ is the Fermi energy for symmetric nuclear matter.

[16] The unit here is 1 Fermi = 1 F $\equiv 10^{-13}$ cm.

[17] Note that nuclear matter is actually rather diffuse. The pion Compton wavelength, which roughly characterizes the range of the nuclear force, is $\hbar/m_{\pi^\pm} c = 1.414\,\text{F}$, while the volume per baryon in nuclear matter is $V/B \approx 5.18\,\text{F}^3$.

(b) The semi-empirical mass formula for the ground-state energy of nuclei contains a term referred to as the *symmetry energy*[18]

$$E_{\rm sym} = a_4 \frac{(N-Z)^2}{B}$$

Show the result in part (a) implies a symmetry energy coefficient of

$$a_4 = \frac{\varepsilon_{\rm F0}}{3}$$

(c) Use the numerical results from Prob. 7.11 to compute a_4. Compare with a measured value of $a_4 = 23.7\,{\rm MeV}$. Discuss.

Solution to Problem 7.12

(a) As in Problem 7.10, we use Eq. (7.116) in the text to relate the number of particles to the maximum wavenumber k_F. We treat neutrons and protons as two separate spin-1/2 Fermi gases and write

$$N = \frac{B(1+\delta)}{2} = \frac{2V}{(2\pi)^3}\frac{4\pi k_{\rm FN}^3}{3}$$

$$Z = \frac{B(1-\delta)}{2} = \frac{2V}{(2\pi)^3}\frac{4\pi k_{\rm FP}^3}{3}$$

which can be inverted to give

$$k_{\rm FN} = \left[3\pi^2 \frac{B}{V}\frac{(1+\delta)}{2}\right]^{1/3}$$

$$k_{\rm FP} = \left[3\pi^2 \frac{B}{V}\frac{(1-\delta)}{2}\right]^{1/3}$$

The total kinetic energy of the system will then be [see Eq. (7.119)][19]

$$T = T_{\rm N} + T_{\rm P} = \frac{3}{5}\frac{B(1+\delta)}{2}\frac{\hbar^2 k_{\rm FN}^2}{2m_p} + \frac{3}{5}\frac{B(1-\delta)}{2}\frac{\hbar^2 k_{\rm FP}^2}{2m_p}$$

$$= \frac{3}{10}\frac{\hbar^2}{2m_p}\left(\frac{3\pi^2}{2}\frac{B}{V}\right)^{2/3} B\left[(1+\delta)^{5/3} + (1-\delta)^{5/3}\right]$$

For $|\delta| \ll 1$ we may expand the factors $(1\pm\delta)^{5/3}$ up to order δ^2, as explained

[18] See [Walecka (2008)].

[19] As in the previous problem, we use m_p for the nucleon mass.

in the footnote in Prob. 7.10, with the result

$$\begin{aligned} T &\approx \frac{3}{10}\frac{\hbar^2}{2m_p}\left(\frac{3\pi^2}{2}\frac{B}{V}\right)^{2/3} B\left[2+\frac{10}{9}\delta^2\right] \\ &= \frac{3}{10}\varepsilon_{\rm F}^0 B\left(2+\frac{10}{9}\delta^2\right) \end{aligned}$$

and therefore

$$\frac{\Delta T}{B} = \frac{\varepsilon_{\rm F0}}{3}\delta^2$$

(b) The symmetry-energy term in the semi-empirical mass formula may be expressed directly in terms of δ as

$$\begin{aligned} E_{\rm sym} &= a_4\frac{(N-Z)^2}{B} \\ &= a_4 B\delta^2 \end{aligned}$$

A comparison with the result of part (a) then yields

$$a_4 = \frac{\varepsilon_{\rm F0}}{3}$$

(c) With the given value of $k_{\rm F0} = 1.42\,{\rm F}^{-1}$ from Prob. 7.11, the Fermi energy of symmetric nuclear matter is

$$\varepsilon_{\rm F0} = \frac{\hbar^2 k_{\rm F0}^2}{2m_p} = 41.8 \text{ MeV}$$

and the result in part (b) then gives

$$a_4 = 13.9\,\text{MeV}$$

This is substantially below the observed value

$$(a_4)_{\rm exp} = 23.7\,\text{MeV}$$

Clearly, although the right idea, attempting to make a quantitative calculation of the symmetry energy without taking into account the nuclear interactions is too naive.

Problem 7.13 Consider an ultra-relativistic Fermi gas at zero temperature. The particle energy is now given by

$$\varepsilon(k) = pc = \hbar kc \qquad ; \; k \equiv |\mathbf{k}|$$

(a) Show the energy per particle in the assembly is

$$\frac{E}{N} = \frac{3}{4}\hbar k_{\mathrm{F}} c = \frac{3}{4}\hbar c \left(\frac{6\pi^2}{g_s}\right)^{1/3} \left(\frac{N}{V}\right)^{1/3}$$

(b) Show the pressure is

$$\begin{aligned} P &= -\left(\frac{\partial E}{\partial V}\right)_N \\ &= \frac{1}{4}\hbar c \left(\frac{6\pi^2}{g_s}\right)^{1/3} \left(\frac{N}{V}\right)^{4/3} \end{aligned}$$

(c) Hence conclude the equation of state is

$$PV = \frac{1}{3}E$$

Solution to Problem 7.13

(a) The number of systems and energy of this relativistic Fermi gas are

$$\begin{aligned} \frac{N}{V} &= \frac{g_s}{(2\pi)^3} \int_0^{k_{\mathrm{F}}} d^3k = \frac{g_s}{6\pi^2} k_{\mathrm{F}}^3 \\ \frac{E}{V} &= \frac{g_s}{(2\pi)^3} \int_0^{k_{\mathrm{F}}} \hbar k c \, d^3k = \frac{g_s}{8\pi^2} \hbar c k_{\mathrm{F}}^4 \end{aligned}$$

Therefore

$$\frac{E}{N} = \frac{3}{4}\hbar k_{\mathrm{F}} c = \frac{3}{4}\hbar c \left(\frac{6\pi^2}{g_s}\right)^{1/3} \left(\frac{N}{V}\right)^{1/3}$$

where the last equality comes from solving N/V for k_{F}.

(b) Since the V dependence is now explicit, the ground-state pressure is readily determined by differentiation

$$\begin{aligned} P &= -\left(\frac{\partial E}{\partial V}\right)_N \\ &= \frac{1}{4}\hbar c \left(\frac{6\pi^2}{g_s}\right)^{1/3} \left(\frac{N}{V}\right)^{4/3} \end{aligned}$$

(c) A comparison of the results in parts (a) and (b) immediately yields the equation of state of this relativistic Fermi gas

$$PV = \frac{1}{3}E$$

Problem 7.14 As a model of a *white-dwarf star*, consider an electically neutral gas composed of a uniform background of inert, fully-ionized He (α particles) and a degenerate (zero-temperature) Fermi gas of electrons.[20]

(a) Show the equation of *local hydrostatic equilibrium* is

$$\frac{1}{\rho}\frac{dP}{dr} = -\frac{4\pi G}{r^2}\int_0^r \rho(r')r'^2dr'$$

where $P(r)$ is the pressure, $\rho(r)$ is the mass density, and G is Newton's gravitational constant. What are the boundary conditions?

(b) Write this equation in the low-density (non-relativistic electron gas) and high-density (relativistic electron gas) limits assuming an ideal Fermi assembly at zero temperature (note Prob. 7.13);

(c) Find expressions for the mass density $\rho(r)$ and the relation between the total mass M and the radius R of the star;

(d) Show that there exists a maximum mass $M_{\rm max}$ comparable with the solar mass $M_\odot$. Explain the physics of why this is so;

(e) Check the initial model using the typical parameters of a white-dwarf: $\rho \approx 10^7\,{\rm g/cm}^3 \approx 10^7\rho_\odot$; $M \approx 10^{33}\,{\rm g} \approx M_\odot$. What is the Fermi energy? Note the following results obtained by numerical integration [Landau and Lifshitz (1980)]:

$$\begin{aligned}
&\frac{1}{\xi^2}\frac{d}{d\xi}\left(\xi^2\frac{df}{d\xi}\right) = -f^{3/2} && ;\ f'(0)=0 \quad ;\ f(1)=0\\
\Longrightarrow \quad & f(0) = 178.2 && ;\ f'(1) = -132.4
\end{aligned}$$

and

$$\begin{aligned}
&\frac{1}{\xi^2}\frac{d}{d\xi}\left(\xi^2\frac{df}{d\xi}\right) = -f^{3} && ;\ f'(0)=0 \quad ;\ f(1)=0\\
\Longrightarrow \quad & f(0) = 6.897 && ;\ f'(1) = -2.018
\end{aligned}$$

Solution to Problem 7.14

(a) We denote the gravitational force per unit mass (the gravitational *field*) in the medium by $\boldsymbol{\mathcal{F}}_{\rm grav}$.[21] Then, in a local-density approximation, the condition of *hydrostatic equilibrium* is that the gravitational attraction toward the center of the star should just be compensated by the repulsive pressure force arising from the degenerate electron gas (see Fig. 7.2).

[20] See, for example, [Walecka (2008)]. This problem is a little longer, but it is well worth it. The electron gas here is treated with Thomas-Fermi theory.

[21] Analogous to the electric field $\boldsymbol{\mathcal{E}}$ in electrodynamics.

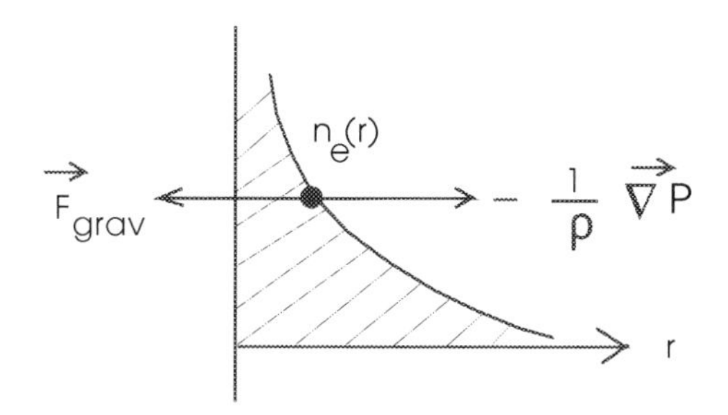

Fig. 7.2 In analogy to the Thomas-Fermi theory of atoms, in white-dwarf stars the attractive (gravitational force)/(unit mass) is balanced by the repulsive pressure force arising from the relativistic, degenerate Fermi gas of electrons.

$$\boldsymbol{\mathcal{F}}_{\text{grav}} = \frac{1}{\rho}\boldsymbol{\nabla}P \qquad ; \text{ hydrostatic equilibrium}$$

The gravitational force points in the radial direction, and exactly as in Thomas-Fermi theory,[22] Gauss's theorem allows one to calculate the gravitational field at a radius r arising from a spherically symmetric mass distribution

$$(\mathcal{F}_{\text{grav}})_r = -\frac{4\pi G}{r^2}\int_0^r \rho(r')r'^2\,dr'$$

Here G is Newton's gravitational constant

$$G = 6.673 \times 10^{-11}\,\text{m}^3/\text{kg-s}^2 \qquad ; \text{ Newton's constant}$$

Thus the condition of hydrostatic equilibrium in the star becomes

$$\frac{1}{\rho}\frac{dP}{dr} = -\frac{4\pi G}{r^2}\int_0^r \rho(r')r'^2\,dr'$$

There are some *boundary conditions* that go along with this equation:

- The mass density clearly vanishes at the surface of the star, which lies at a radius R

$$\rho(R) = 0 \qquad ; \text{ surface}$$

- At the origin, the gravitational force will vanish, while the mass density will be finite. Hence, from above, the slope of the pressure will vanish

[22]See Prob. 7.25.

there[23]

$$\frac{dP(0)}{dr} = 0 \qquad ; \ \rho(0) \neq 0$$

A white-dwarf star has the following properties:

- Totally ionized helium atoms $^4\text{He}^{++}$ (α-particles) provide an inert, positive, uniformly-charged background in which the electrons move;
- The *mass* of the star comes from the $^4\text{He}^{++}$;
- The electrons provide the repulsive *pressure* that keeps the star from collapsing under the gravitational attraction.

Since the entire medium is neutral, the α-particle number density n_α is related to the electron density n_e by

$$n_\alpha = \frac{1}{2} n_e \qquad ; \text{ neutrality}$$

Since the mass comes from the α-particles, the mass density ρ is given by

$$\rho = m_{\text{He}} n_\alpha \approx 4 m_p n_\alpha = 2 m_p n_e \qquad ; \text{ mass density}$$

(b) We will do the more interesting extreme relativistic limit (ERL) here.[24] From Prob. 7.13, the pressure in a degenerate relativistic Fermi gas of electrons is given by

$$P = \frac{1}{4} \hbar c \left(\frac{6\pi^2}{g_s} \right)^{1/3} n^{4/3} \qquad ; \text{ relativistic F-G}$$

where $n = N/V$ is the number density. The equation of state to be employed follows as

$$P = \eta \rho^{4/3} \qquad ; \text{ equation of state}$$

$$\eta = \frac{\hbar c}{4} \left(\frac{6\pi^2}{g_s} \right)^{1/3} \frac{1}{(2m_p)^{4/3}}$$

Insertion into the relation in part (a) gives

$$\frac{4\eta}{3} \frac{1}{\rho^{2/3}} \frac{d\rho}{dr} = 4\eta \frac{d}{dr} \rho^{1/3} = -\frac{4\pi G}{r^2} \int_0^r \rho(r') r'^2 \, dr'$$

---

[23] There is no *cusp* in the pressure at the origin.

[24] The corresponding calculation of the properties of a white-dwarf star in the non-relativistic limit (NRL) appropriate to small mass stars, and the interpolation between the NRL and ERL, are discussed in detail in Probs. 8.16 and 8.18 in [Amore and Walecka (2013)].

Introduce the definitions

$$\xi = \frac{r}{R} \qquad ; \; R = \text{ star radius}$$

$$\bar{\lambda}^2 = \frac{4\pi G R^2}{4\eta}$$

The above equation then takes the form

$$\xi^2 \frac{d}{d\xi}\rho^{1/3} = -\bar{\lambda}^2 \int_0^{\xi} \rho(\xi')\xi'^2 \, d\xi'$$

This can now be converted into dimensionless form with the introduction of[25]

$$\bar{\lambda}^3 \rho \equiv f^3 \qquad ; \; \text{dimensionless}$$

Differentiation with respect to ξ then recasts the condition of hydrostatic equilibrium into the following

$$\frac{1}{\xi^2}\frac{d}{d\xi}\left(\xi^2 \frac{d}{d\xi} f\right) = -f^3 \qquad ; \; \text{N-L diff eqn}$$

This dimensionless, non-linear differential equation for the density is to be solved with the two boundary conditions in part (a), which can be written as

$$f(1) = 0 \quad ; \; f'(0) = 0 \qquad ; \; \text{B.C.}$$

(c) The total mass of the star is obtained by integrating the mass density out to the star's radius

$$M = 4\pi \int_0^R \rho(r) r^2 \, dr = 4\pi R^3 \int_0^1 \rho(\xi)\xi^2 \, d\xi$$

$$= \frac{4\pi R^3}{\bar{\lambda}^3} \int_0^1 f^3 \xi^2 \, d\xi$$

The differential equation for f can now be used to substitute for f^3 in the integral

$$\int_0^1 f^3 \xi^2 \, d\xi = -\int_0^1 \frac{d}{d\xi}\left(\xi^2 \frac{d}{d\xi} f\right) d\xi = -f'(1)$$

[25] The dimensions of $\eta/\hbar c$ are $[M^{-4/3}]$, of $G/\hbar c$ are $[M^{-2}]$, and of $\bar{\lambda}^3$ are $[M^{-1}L^3]$.

Hence

$$M = \frac{4\pi R^3}{\bar{\lambda}^3}[-f'(1)] = 4\pi R^3 \left[\frac{4\eta}{4\pi G R^2}\right]^{3/2}[-f'(1)]$$

Note that the radius of the star R cancels in this expression, and it becomes

$$M = -4\pi\left(\frac{\hbar c}{4\pi G}\right)^{3/2}\left(\frac{6\pi^2}{g_s}\right)^{1/2}\frac{1}{(2m_p)^2}f'(1)$$

The problem has thus been reduced to integrating the nonlinear differential equation for f subject to the two given boundary conditions, and determining the resulting $f'(1)$, which then provides the mass of the star through the above.

(d) The results obtained from numerical integration using Mathcad 11 and the Runge-Kutta algorithm are shown in Fig. 7.3. The differential

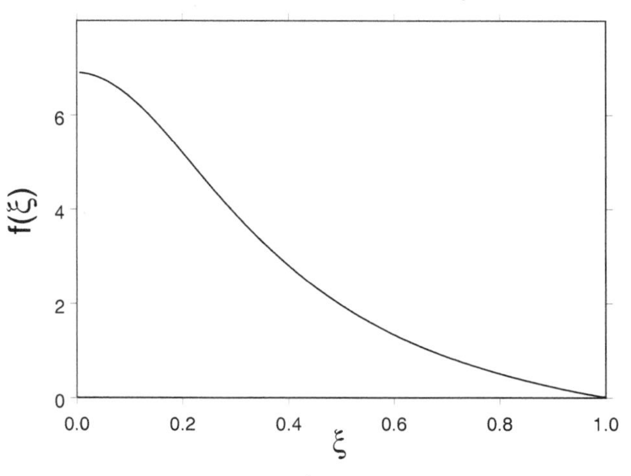

Fig. 7.3 Dimensionless white-dwarf mass density $f(\xi) = \bar{\lambda}^3\rho$ versus $\xi = r/R$. Here $\bar{\lambda}^3 = (2m_p)^2(4\pi G/\hbar c)^{3/2}(g_s/6\pi^2)^{1/2}R^3$. Taken from [Walecka (2008)].

equation was integrated in from $\xi = 1$, and the initial slope was adjusted until the curve became flat at the origin. The values obtained (rounded to three decimal places) are

$$f'(1) = -2.018 \qquad ; \text{ numerical integration}$$
$$f(0) = 6.897$$

The total mass of the white-dwarf can now be calculated from the above, using $g_s = 2$ and the following numerical values (in m.k.s.)

$$G = 6.67 \times 10^{-11}\,\text{Nm}^2/\text{kg}^2 \qquad ; \; c = 3.00 \times 10^8\,\text{m/s}$$
$$m_p = 1.67 \times 10^{-27}\,\text{kg} \qquad ; \; \hbar = 1.05 \times 10^{-34}\,\text{J-s}$$

The result is

$$M = 2.85 \times 10^{30}\,\text{kg}$$

It is instructive to write the total mass in terms of the mass of the sun

$$M_{\odot} = 1.99 \times 10^{30}\,\text{kg} \qquad ; \text{ solar mass}$$

It follows that

$$\frac{M}{M_{\odot}} = 1.43 \qquad ; \text{ Chandrasekhar limit}$$

This is known as the *Chandrasekhar limit* for the mass of a white-dwarf. This is the largest mass that can be supported against the gravitational attraction by the Fermi pressure of a cold, fully-relativistic, electron gas.

If the star has a larger mass than the Chandrasekhar limit, it collapses down to densities where the *nuclear force* comes into play, with resulting supernovae and neutron stars. Eventually, if the mass is large enough, black holes are formed.

(e) The mass of the white dwarf comes from the helium, with an atomic mass of $m_{\text{He}} = 6.64 \times 10^{-24}\,\text{gm}$ (see Prob. 2.8). Thus at a mass density of $\rho \approx 10^7\,\text{gm/cm}^3$, the number densities in the white dwarf satisfy

$$n_{\text{e}} = 2n_{\text{He}} = 2\rho/m_{\text{He}} \approx 3.01 \times 10^{30}\,\text{cm}^{-3}$$

As an estimate, we calculate the Fermi energy in the ERL of part (b) and Prob. 7.13[26]

$$\varepsilon_{\text{F}} = \hbar k_{\text{F}} c = \hbar c (3\pi^2 n_{\text{e}})^{1/3}$$
$$\approx 8.81 \times 10^5\,\text{eV}$$

This gives an equivalent Fermi temperature of

$$T_{\text{F}} = \varepsilon_{\text{F}}/k_{\text{B}} \approx 1.02 \times 10^{10}\,^{\text{o}}\text{K}$$

As this is very far above the temperature of white dwarf stars, we are justified in treating the electrons as a degenerate Fermi gas.

---

[26]Use $\hbar c = 1.973 \times 10^{-5}$ eV-cm, and $k_{\text{B}} = 8.620 \times 10^{-5}\,\text{eV}/^{\text{o}}\text{K}$ (see Prob. 3.20).

Problem 7.15 Liquid hydrogen has a mass density of $\rho_{\rm H_2} = .07\,{\rm g/cm^3}$. Suppose that at a density of $\sim 1\,{\rm g/cm^3}$ it were to go into a metallic state. Compute the Fermi energy of that metal in eV.

Solution to Problem 7.15

We first calculate the relevant number density. From Prob. 2.8, the proton mass is

$$\begin{aligned} m_p &= 1.673 \times 10^{-24}\,{\rm gm} \\ \hbar^2/2m_p &= 20.74 \times 10^{-20}\,{\rm eV\text{-}cm^2} \end{aligned}$$

With a mass density of $\rho \approx 1\,{\rm g/cm^3}$ of hydrogen, the proton number density is

$$n_p \approx \frac{\rho}{m_p} = 5.98 \times 10^{23}\,{\rm cm^{-3}}$$

The proton Fermi energy follows from Eqs. (7.113) and (7.116)

$$\varepsilon_{\rm F} = \frac{\hbar^2 k_{\rm F}^2}{2m_p} = \frac{\hbar^2}{2m_p}\left(\frac{6\pi^2 n_p}{g_s}\right)^{2/3}$$

Here $g_s = 2$. Therefore, in the metal the proton Fermi energy would be

$$\varepsilon_{\rm F} = 1.41 \times 10^{-2}\,{\rm eV} \qquad ;\ {\rm proton}$$

The electron mass is

$$m_e = 9.111 \times 10^{-28}\,{\rm gm} \qquad ;\ m_p/m_e = 1836$$

The metal is neutral, so $n_e = n_p$. Therefore, the electron Fermi energy would be m_p/m_e larger

$$\varepsilon_{\rm F} = 25.9\,{\rm eV} \qquad ;\ {\rm electron}$$

Note that this is roughly twice the binding energy of the hydrogen atom.

Problem 7.16 When a metal is heated to a sufficiently high temperature, electrons are emitted from the metal surface and can be collected as thermionic current. Assume the electrons form a non-interacting Fermi gas, and derive the Richardson-Dushman equation for the current

$$i = \frac{4\pi e m (k_{\rm B}T)^2}{h^3} e^{-W/k_{\rm B}T}$$

where W is the work function for the metal (that is, the energy necessary to remove the electrons).

Solution to Problem 7.16

The work function W is the energy required to eject an electron from the metal at $T = 0$. Thus those electrons that get out must have (see Fig. 7.4)

$$\varepsilon_x - \varepsilon_{\rm F} \geq W$$

where we assume[27]

$$\varepsilon_{\rm F} \approx \mu$$

We are working way out in the tail of the Fermi distribution, and there must be enough energy of motion in the x-direction to get over the wall.

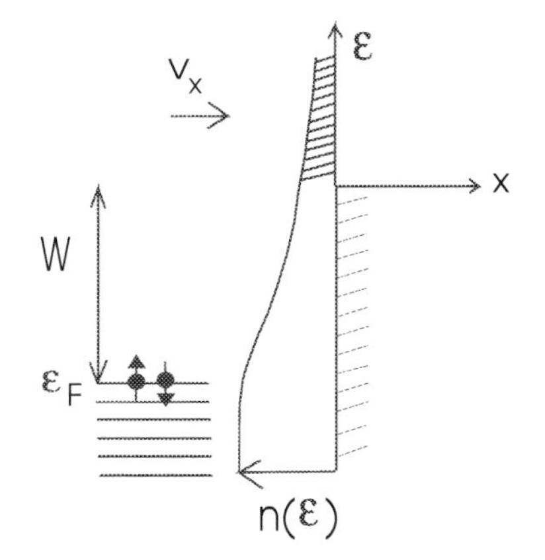

Fig. 7.4 Escaping electrons for the thermionic effect in a metal.

The current in the x-direction is

$$j_x = en\langle fv_x \rangle$$

Here n is the electron density, and $\langle fv_x \rangle$ is the fraction of contributing electrons times the x-component of their velocity given by

$$\langle fv_x \rangle = \frac{1}{N}\frac{g_s V}{(2\pi\hbar)^3}\int\cdots\int \frac{1}{e^{(\varepsilon-\mu)/k_{\rm B}T}+1}\left(\frac{p_x}{m}\right)dp_x dp_y dp_z \qquad ; \; \varepsilon_x - \mu \geq W$$

[27]See Fig. 7.9 in the text; here we measure energies from the bottom of the well.

where we only integrate over $p_x \geq 0$ to get the positive flux.

If we assume that $W/k_{\rm B}T \gg 1$, then the Fermi distribution becomes

$$\frac{1}{e^{(\varepsilon-\mu)/k_{\rm B}T}+1} \approx e^{(\mu-\varepsilon)/k_{\rm B}T} \qquad ; \frac{W}{k_{\rm B}T} \gg 1$$

The above integral then takes the form

$$\begin{aligned}\mathcal{I} &\equiv \int \cdots \int e^{(\mu-\varepsilon)/k_{\rm B}T}\left(\frac{p_x}{m}\right) dp_x dp_y dp_z \\ &= \int_{-\infty}^{\infty} e^{-p_y^2/2mk_{\rm B}T}\, dp_y \int_{-\infty}^{\infty} e^{-p_z^2/2mk_{\rm B}T}\, dp_z \int e^{(\mu-\varepsilon_x)/k_{\rm B}T}\, \frac{p_x\, dp_x}{m}\end{aligned}$$

This is to be evaluted subject to the condition

$$\varepsilon_x \geq W + \mu \qquad ; \varepsilon_x = \frac{p_x^2}{2m}$$

Now use

$$d\varepsilon_x = \frac{p_x\, dp_x}{m}$$

Then $\mathcal{I}$ becomes[28]

$$\mathcal{I} = (2\pi m k_{\rm B}T)\int_{W+\mu}^{\infty} d\varepsilon_x\, e^{(\mu-\varepsilon_x)/k_{\rm B}T}$$

With $u \equiv (\varepsilon_x - \mu)/k_{\rm B}T$, it follows that

$$\mathcal{I} = (2\pi m k_{\rm B}T)k_{\rm B}T\int_{W/k_{\rm B}T}^{\infty} e^{-u}\, du = (2\pi m k_{\rm B}T)k_{\rm B}T\, e^{-W/k_{\rm B}T}$$

Identify the electron density as

$$n \equiv \frac{N}{V}$$

Thus, from the above

$$ne\langle f v_x\rangle = \frac{e g_s}{(2\pi\hbar)^3}(2\pi m k_{\rm B}T)k_{\rm B}T\, e^{-W/k_{\rm B}T}$$

Hence, in summary[29]

$$i = \frac{4\pi e m (k_{\rm B}T)^2}{h^3} e^{-W/k_{\rm B}T}$$

---

[28] Note from Eq. (2.127) $\int_{-\infty}^{\infty} dx\, e^{-x^2} = \sqrt{\pi}$.

[29] We use $g_s = 2$, and $2\pi\hbar = h$.

where $W \gg k_{\rm B}T$ is the work function for the metal (that is, the energy necessary to remove the electrons at $T = 0$). This is the Richardson-Dushman equation for the thermionic current.

Problem 7.17 Prove the *Bohr-Van Leeuwen theorem*, which states that the magnetic susceptibility of an assembly of charged point particles obeying classical mechanics and classical statistics *vanishes identically.* Introduce the magnetic field by means of a vector potential so that only the kinetic energy contains the magnetic field in the form

$$\mathcal{T} = \frac{1}{2m}\left[\mathbf{p} - \frac{e}{c}\mathbf{A}(\mathbf{x})\right]^2$$

(a) First prove the result with $H = \mathcal{T}$;

(b) Then show that the result holds even in the presence of two-body interactions, where $H = \mathcal{T} + \sum_{i<j\leq N} V(ij)$.

Solution to Problem 7.17

(a) First, choose a gauge that relates the vector potential to the magnetic field as in Eq. (7.179)

$$\begin{aligned}\mathbf{A}(\mathbf{x}) &= [-yB, 0, 0] \qquad ; \text{ choice of gauge}\\ \boldsymbol{\nabla}\times\mathbf{A} &= [0, 0, B]\end{aligned}$$

Then start in the microcanonical ensemble, where the one-particle kinetic energy is that in Eq. (7.181)

$$\mathcal{T} = \frac{1}{2m}\left[\mathbf{p} - \frac{e}{c}\mathbf{A}(\mathbf{x})\right]^2$$

The classical partition function is given by

$$(\text{p.f.})_{\rm cl} = \frac{1}{h^3}\int d^3x \int d^3p\, e^{-[\mathbf{p}-(e/c)\mathbf{A}(\mathbf{x})]^2/2mk_{\rm B}T}$$

Now fix $\mathbf{x}$, and change variables in the momentum integral to $\mathbf{p}'$ where

$$\begin{aligned}\mathbf{p}' &\equiv \mathbf{p} - \frac{e}{c}\mathbf{A}(\mathbf{x})\\ d^3p' &= d^3p\end{aligned}$$

The partition function then becomes

$$(\text{p.f.})_{\rm cl} = \frac{1}{h^3}\int d^3x \int d^3p'\, e^{-\mathbf{p}'^2/2mk_{\rm B}T}$$

The field has *disappeared* from this expression! The corresponding classical Helmoltz free energy is given by Eq. (2.107)

$$A(T,V,N,B) = -k_{\rm B}T \ln \frac{({\rm p.f.})_{\rm cl}^N}{N!}$$

and the magnetization by Eq. (3.179)

$$\mathcal{M} = -\left(\frac{\partial A}{\partial B}\right)_{T,V,N}$$

This clearly vanishes.

(b) Now include an interaction term, assumed to be a function of the inter-particle distances, so that the total hamiltonian becomes

$$H = \sum_{i=1}^{N} \frac{1}{2m}\left[\mathbf{p}_i - \frac{e}{c}\mathbf{A}(\mathbf{x}_i)\right]^2 + \sum_{i<j\leq N} V(|\mathbf{x}_i - \mathbf{x}_j|)$$

Work in the canonical ensemble, where the classical partition function is given in Eq. (4.50)

$$({\rm P.F.})_{\rm cl} = \frac{1}{N!\,h^{3N}} \int \cdots \int d^3x_1 \cdots d^3x_N \int \cdots \int d^3p_1 \cdots d^3p_N \, e^{-H/k_{\rm B}T}$$

Again, fix the coordinates and do the integrals over momenta. The terms in $e^{-\mathcal{T}/k_{\rm B}T}$ all factor, as does $e^{-V/k_{\rm B}T}$ which becomes part of the spatial integrals. Each momentum integral can then be handled with a change of variables as in part (a). The partition function thus becomes

$$({\rm P.F.})_{\rm cl} = \frac{1}{N!\,h^{3N}} \int \cdots \int d^3x_1 \cdots d^3x_N \int \cdots \int d^3p'_1 \cdots d^3p'_N \, e^{-H'/k_{\rm B}T}$$

$$H' = \sum_{i=1}^{N} \frac{1}{2m}\mathbf{p}_i'^2 + \sum_{i<j\leq N} V(|\mathbf{x}_i - \mathbf{x}_j|)$$

The field has again *disappeared* from the partition function. The Helmholtz free energy is given by Eq. (4.47)

$$A(T,V,N,B) = -k_{\rm B}T \ln ({\rm P.F.})_{\rm cl}$$

and the magnetization again vanishes as above. This is the *Bohr-Van Leeuwen theorem.*

Problem 7.18 (a) Show that B_0 in Eq. (7.234) has the dimensions of a magnetic field;

(b) If $\varepsilon_{\rm F}^0$ is the free Fermi energy and μ_0 the appropriate Bohr magneton, show the following ratio is a pure number

$$\frac{|\mu_0|B_0}{\varepsilon_{\rm F}^0} = \frac{1}{3}\left(\frac{3}{2}\right)^{1/3}$$

Solution to Problem 7.18

(a) The quantity B_0 in Eq. (7.234) is given by

$$B_0 = \frac{\pi\hbar c}{|e|}\left(\frac{1}{2\pi g_s}\right)^{2/3}\left(\frac{N}{V}\right)^{2/3}$$

Let μ_0 be the appropriate Bohr magneton [compare Eq. (7.144)]

$$\mu_0 = \frac{e\hbar}{2mc}$$

To show that B_0 has the dimensions of a magnetic field, we shall show $|\mu_0|B_0$ has the dimensions of energy. Consider

$$|\mu_0|B_0 = \frac{\hbar^2\pi^2}{m}\left(\frac{1}{2\pi g_s}\right)^{2/3}\left(\frac{N}{V}\right)^{2/3}$$

Since $(N/V)^{2/3}$ has dimension $[L^{-2}]$, it is clear from Eq. (2.113) that this expression has the dimension of energy.[30]

(b) From Eqs. (7.113) and (7.116), the Fermi energy is

$$\varepsilon_{\rm F}^0 = \frac{\hbar^2 k_{\rm F}^2}{2m} = \frac{\hbar^2}{2m}\left[\left(\frac{6\pi^2}{g_s}\right)\left(\frac{N}{V}\right)\right]^{2/3} \qquad ; \quad \frac{N}{V} = \frac{g_s k_{\rm F}^3}{6\pi^2}$$

Hence the dimensionless ratio

$$\frac{|\mu_0|B_0}{\varepsilon_{\rm F}^0} = \frac{2}{(12)^{2/3}} = \frac{1}{3}\left(\frac{3}{2}\right)^{1/3}$$

Problem 7.19 (a) Reproduce the numerical results in Figs. 7.15 and 7.16 in the text;

(b) Extend these results in both directions.

[30] See also the dimensionless result in part (b); recall $h = 2\pi\hbar$.

Solution to Problem 7.19

In Figures 7.5 and 7.6 we display a re-calculation of Figs. 7.15–7.16 in the text describing Landau diamagnetism, which extends the range of the calculation slightly in both directions.[31]

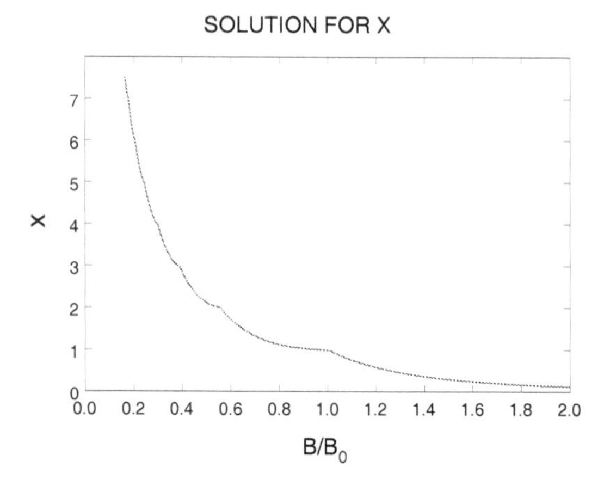

Fig. 7.5 Numerical solution of the first of Eqs. (7.233) for x as a function of B/B_0. Note the breaks in the curve at integer values of x.

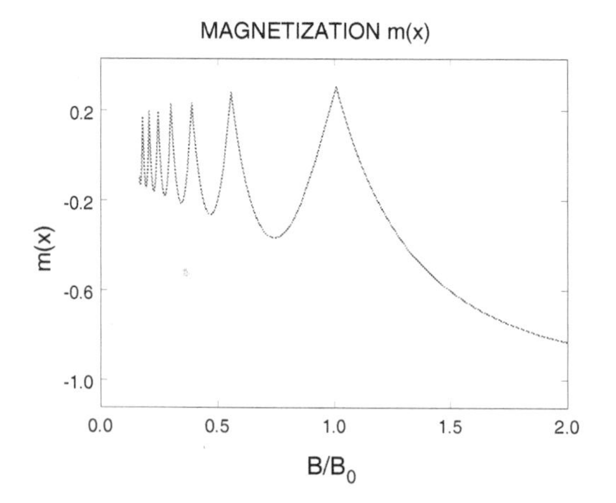

Fig. 7.6 Calculation of the magnetization $m(x)$ based on Fig. 7.5 and the second of Eqs. (7.233). Here $\mathcal{M}/V \equiv (|e|\hbar/2mc)(N/V)m(x)$.

[31] The calculation uses Mathcad 11. The extension is limited only by our ability to plot the results.

Problem 7.20 Show by direct differentiation that the magnetization $\mathcal{M}$ for Pauli spin paramagnetism in Eq. (7.153) is given by

$$\mathcal{M} = k_{\rm B}T\frac{\partial}{\partial B}\ln{\rm (G.P.F.)}$$

where (G.P.F.)(μ, V, T, B) is given by Eq. (7.151).

Solution to Problem 7.20

The grand partition function for an assembly of spin-1/2 systems in a magnetic field B is given in Eq. (7.151)

$$({\rm G.P.F.})_B = \prod_i \left[1 + e^{(\mu-\varepsilon_i+\mu_0 B)/k_{\rm B}T}\right]\left[1 + e^{(\mu-\varepsilon_i-\mu_0 B)/k_{\rm B}T}\right]$$

Here the first factor arise from spin-up, and the second from spin-down. This is a function of (V, T, μ, B).[32] Fix the other three variables, and differentiate the ln of this expression with repect to B

$$\begin{aligned} k_{\rm B}T\frac{\partial}{\partial B}\ln{\rm (G.P.F.)}_B &= \mu_0 \sum_i \left[\frac{1}{e^{(\varepsilon_i-\mu-\mu_0 B)/k_{\rm B}T}+1} - \frac{1}{e^{(\varepsilon_i-\mu+\mu_0 B)/k_{\rm B}T}+1}\right] \\ &= \mu_0 \sum_i (n_{i\uparrow} - n_{i\downarrow}) \end{aligned}$$

This is the expression for the magnetization in Eq. (7.153). Hence

$$\mathcal{M} = k_{\rm B}T\frac{\partial}{\partial B}\ln{\rm (G.P.F.)}_B$$

Problem 7.21 The expressions for the simple harmonic oscillator wave functions $\psi_n(\xi)$ can be found in [Schiff (1968)]

$$\psi_n(\xi) = \left(\frac{1}{\sqrt{\pi}\,2^n n!}\right)^{1/2}\left[(-1)^n e^{\xi^2}\frac{\partial^n}{\partial\xi^n}e^{-\xi^2}\right]e^{-\xi^2/2}$$

$$\int_{-\infty}^{\infty}|\psi_n(\xi)|^2 d\xi = 1$$

Plot the probability distribution in the relative coordinate $y - y_0$ [see Eq. (7.196)] for the orbits with several values of n for a charged particle in a uniform magnetic field.

[32]Remember μ is the chemical potential.

Solution to Problem 7.21

In Figure 7.7 we plot the probability densities

$$\rho_n(\xi) \equiv |\psi_n(\xi)|^2$$
$$\xi = \left(\frac{|e|B_0}{\hbar c}\right)^{1/2}(y - y_0) \equiv \frac{1}{\alpha}(y - y_0)$$

for the first four orbits for a charged particle in a uniform magnetic field.

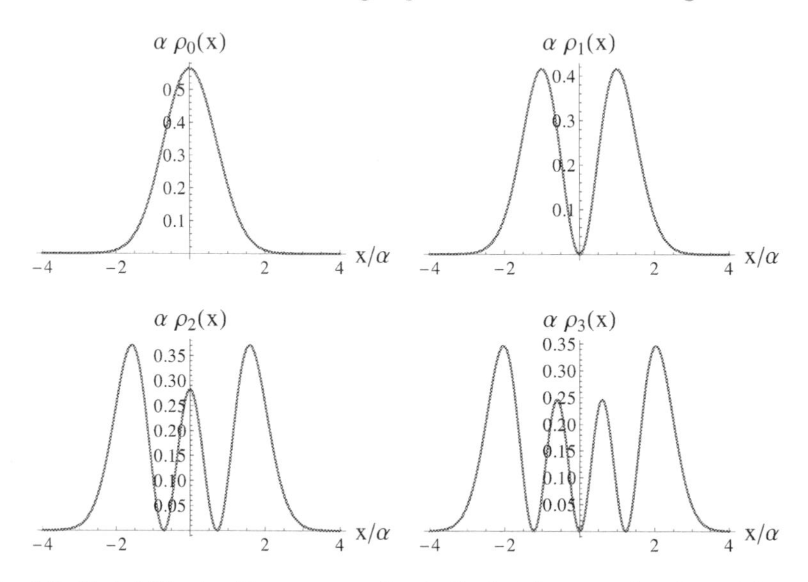

Fig. 7.7 Probability densities corresponding to the first four wave functions of the one-dimensional simple harmonic oscillator, where $x \equiv y - y_0$, and $\alpha \equiv (\hbar c/|e|B_0)^{1/2}$. Taken from [Amore and Walecka (2013)].

Problem 7.22 Consider a two-dimensional non-relativistic Fermi gas in a large square with side L and periodic boundary conditions.

(a) Show that as $L \to \infty$, the sum over states of a function of $|\mathbf{k}|$ becomes

$$\sum_i \to g_s\left(\frac{L}{2\pi}\right)^2 \int_{\text{all k}} d^2k = g_s\frac{L^2}{2\pi}\int_0^\infty k\,dk = g_s\frac{\mathcal{A}}{4\pi}\left(\frac{2m}{\hbar^2}\right)\int_0^\infty d\varepsilon$$

where $\mathcal{A} = L^2$ is the area, and $\varepsilon(k) = \hbar^2k^2/2m$;

(b) Assume the Fermi gas is at temperature $T = 0$. Show

$$\frac{N}{\mathcal{A}} = \frac{g_s k_{\rm F}^2}{4\pi} \qquad ; \quad \frac{E}{N} = \frac{1}{2}\varepsilon_{\rm F}$$

where the Fermi energy is $\varepsilon_{\rm F} = \hbar^2 k_{\rm F}^2/2m$.

Solution to Problem 7.22

(a) The results in Probs. 5.1–5.2 are immediately reduced down from three to two dimensions

$$\sum_{n_x}\sum_{n_y} f(\mathbf{k}^2) \rightarrow \left(\frac{L}{2\pi}\right)^2 \int_{\text{all k}} f(k^2)\, d^2k \qquad ; \; L \rightarrow \infty$$
$$= \left(\frac{L}{2\pi}\right)^2 \int_0^\infty f(k^2)\, 2\pi k\, dk$$

If a factor g_s is included for the degeneracy of the levels, then for a non-relativistic gas in two dimensions with energy $\varepsilon(k) = \hbar^2 k^2/2m$

$$\sum_i \rightarrow g_s \left(\frac{L}{2\pi}\right)^2 \int_{\text{all k}} d^2k = g_s \frac{L^2}{2\pi} \int_0^\infty k\, dk = g_s \frac{\mathcal{A}}{4\pi}\left(\frac{2m}{\hbar^2}\right) \int_0^\infty d\varepsilon$$

where $\mathcal{A} = L^2$ is the area.

(b) For the non-relativistic Fermi gas at temperature zero, the number of systems and energy are given by filling the levels to $k_{\rm F}$[33]

$$\frac{N}{\mathcal{A}} = \frac{g_s}{(2\pi)^2} \int_0^{k_{\rm F}} 2\pi k\, dk = \frac{g_s k_{\rm F}^2}{4\pi}$$
$$\frac{E}{\mathcal{A}} = \frac{g_s}{(2\pi)^2} \frac{\hbar^2}{2m} \int_0^{k_{\rm F}} 2\pi k^3\, dk = \frac{g_s k_{\rm F}^2}{8\pi} \frac{\hbar^2 k_{\rm F}^2}{2m}$$

It follows that in two dimensions, the energy per particle in the degenerate Fermi gas is one-half the Fermi energy

$$\frac{E}{N} = \frac{1}{2}\varepsilon_{\rm F} \qquad ; \; \varepsilon_{\rm F} = \frac{\hbar^2 k_{\rm F}^2}{2m}$$

Problem 7.23 In a two-dimensional sample, in analogy to the surface tension,[34] the pressure P becomes the *normal force per unit length*.

(a) Show all the thermodynamic arguments in the text go through with the substitution $V \rightarrow \mathcal{A}$, where $\mathcal{A}$ is the area of the sample;

[33] In two dimensions, one has a circular "Fermi disc".

[34] See, for example, [Fetter and Walecka (2003a)]. Here we assume there is a *pressure* and not a *tension* (compare Prob. 6.5).

(b) Show the pressure in Prob. 7.22(b) is

$$P = -\left(\frac{\partial E}{\partial \mathcal{A}}\right)_N = \frac{2\pi}{g_s}\frac{\hbar^2}{2m}\left(\frac{N}{\mathcal{A}}\right)^2$$

(c) Show the finite temperature analogs of Eqs. (7.108) are

$$\frac{N}{\mathcal{A}} = \frac{g_s}{4\pi}\left(\frac{2m}{\hbar^2}\right)\int_0^\infty \frac{d\varepsilon}{e^{(\varepsilon-\mu)/k_BT}+1}$$

$$P = \frac{E}{\mathcal{A}} = \frac{g_s}{4\pi}\left(\frac{2m}{\hbar^2}\right)\int_0^\infty \frac{\varepsilon\, d\varepsilon}{e^{(\varepsilon-\mu)/k_BT}+1}$$

Solution to Problem 7.23

(a) If P is the normal force per unit length in two dimensions, then the argument in Prob. 1.6 and Fig. 1.1 indicates that the work done by the assembly on a wire of length l is

$$dW = Fdx = Pldx = Pd\mathcal{A} \qquad ; \text{ external work}$$

where $ldx = d\mathcal{A}$ is the change in area. The first and second law then become for a reversible quasistatic process

$$dE = TdS - Pd\mathcal{A} \qquad ; \text{ first and second laws}$$

Hence all the thermodynamic arguments go through in two dimensions with the replacement

$$PdV \to Pd\mathcal{A}$$

(b) Solve the first of the expressions in Prob. 7.22(b) for k_F^2, and substitute it in the second

$$k_F^2 = \frac{4\pi}{g_s}\left(\frac{N}{\mathcal{A}}\right)$$

$$E = \frac{2\pi N}{g_s}\frac{\hbar^2}{2m}\left(\frac{N}{\mathcal{A}}\right)$$

Now that the $\mathcal{A}$ dependence is explicit, one can calculate the ground-state pressure from the thermodynamic relation in part (a)[35]

$$P = -\left(\frac{\partial E}{\partial \mathcal{A}}\right)_{N,S} = \frac{2\pi}{g_s}\frac{\hbar^2}{2m}\left(\frac{N}{\mathcal{A}}\right)^2$$

[35]Note that the change in entropy $dS = 0$ if the system remains in the ground state.

Thus for the two-dimensional Fermi gas

$$P = \frac{E}{\mathcal{A}}$$

(c) From Prob. 7.22(a), the sum over states in two dimensions takes the form

$$\sum_i \to g_s \frac{\mathcal{A}}{4\pi}\left(\frac{2m}{\hbar^2}\right)\int_0^\infty d\varepsilon$$

Hence, the two-dimensional analogs of Eqs. (7.108) are

$$\frac{N}{\mathcal{A}} = \frac{g_s}{4\pi}\left(\frac{2m}{\hbar^2}\right)\int_0^\infty \frac{d\varepsilon}{e^{(\varepsilon-\mu)/k_\mathrm{B}T}+1}$$

$$P = \frac{E}{\mathcal{A}} = \frac{g_s}{4\pi}\left(\frac{2m}{\hbar^2}\right)\int_0^\infty \frac{\varepsilon\, d\varepsilon}{e^{(\varepsilon-\mu)/k_\mathrm{B}T}+1}$$

Problem 7.24 A positive point charge Ze_p placed into a uniform electron gas (imposed on a uniform, positive fixed background of charge density $e_p n_0$ that makes the unperturbed system neutral) will be *screened.* The Thomas-Fermi theory of this screening is achieved as follows:[36]

(a) Show the condition of local hydrostatic equilibrium for the electrons is

$$e_p\mathbf{E} = -\frac{1}{n}\boldsymbol{\nabla}P$$

where P is the pressure, $\mathbf{E}$ is the electric field, and $n = N/V$ is the electron density;

(b) Show Poisson's equation for the electric field gives

$$\boldsymbol{\nabla}\cdot\mathbf{E} = -\boldsymbol{\nabla}^2\Phi = 4\pi\left[Ze_p\delta^{(3)}(\mathbf{x}) - e_p(n-n_0)\right]$$

where Φ is the electrostatic potential, with $\mathbf{E} = -\boldsymbol{\nabla}\Phi$;

(c) Write $n - n_0 \equiv \delta n$, and use the last of Eqs. (7.122) for the zero-temperature pressure of the Fermi gas of electrons to show from (a) that

$$\frac{2}{3}\frac{\hbar^2}{2m}\left(3\pi^2\right)^{2/3}\frac{1}{n^{1/3}}\boldsymbol{\nabla}\delta n = e_p\boldsymbol{\nabla}\Phi$$

[36] Problems 7.24–7.25 are long, but invaluable, and the steps are clearly laid out.

Since the l.h.s. is already linear in small quantities, use the first of Eqs. (7.122) to write this as

$$\frac{2}{3}\frac{\hbar^2 k_F^2}{2m}\frac{1}{n_0}\nabla \delta n = e_p \nabla \Phi$$

(d) Take the divergence of this result, and use part (b) to show[37]

$$\left(\nabla^2 - q_{TF}^2\right)\delta n(\mathbf{x}) = -Zq_{TF}^2 \delta^{(3)}(\mathbf{x}) \qquad ; \; q_{TF} \equiv \left(\frac{6\pi e^2 n_0}{\varepsilon_F}\right)^{1/2}$$

(e) Show the solution to this equation gives the Thomas-Fermi result for the induced screening of a point charge in an electron gas (here $r = |\mathbf{x}|$)

$$\delta\rho_{TF}(r) = -Ze_p q_{TF}^2 \frac{e^{-q_{TF}\,r}}{4\pi r} \qquad ; \; \delta\rho = -e_p \delta n$$

(f) Show that the integrated induced density completely screens the point charge.

Solution to Problem 7.24

(a) Consider a small cubic volume with the pressure varying in the x-direction. The pressure force on this volume is

$$F_x = P(x)\,dydz - P(x+dx)\,dydz = -\frac{\partial P}{\partial x}dxdydz$$

This is immediately generalized to

$$\mathbf{F}_P = -\nabla P\,dv \qquad ; \text{ pressure force}$$

If there are ndv electrons in this volume and an electric field $\mathbf{E}$, the electric force is $-e_p\mathbf{E}\,ndv$. In equilibrium, the forces must balance, so

$$-e_p\mathbf{E}\,ndv - \nabla P dv = 0$$

It follows that the forces satisfy

$$e_p\mathbf{E} = -\frac{1}{n}\nabla P$$

(b) Poisson's equation for the electric field is[38]

$$\nabla \cdot \mathbf{E} = -\nabla^2\Phi = 4\pi\rho$$

[37] We correct a misprint in the text; the final exponent is 1/2.

[38] Remember we are using c.g.s. units.

where Φ is the electrostatic potential, with $\mathbf{E} = -\boldsymbol{\nabla}\Phi$. Here ρ is the charge density given by

$$\rho = Ze_p\delta^{(3)}(\mathbf{x}) - e_p(n - n_0)$$

The first term is the density from the inserted charge Ze_p placed the origin, the second term $-e_pn$ is the charge density of the surrounding electrons, and the last term e_pn_0 is that of the neutralizing uniform background charge. Thus

$$\boldsymbol{\nabla}\cdot\mathbf{E} = -\boldsymbol{\nabla}^2\Phi = 4\pi e_p\left[Z\delta^{(3)}(\mathbf{x}) - (n - n_0)\right]$$

(c) The last of Eqs. (7.122) expresses the pressure of the degenerate electron gas in terms of its density[39]

$$P = \frac{2}{5}\frac{\hbar^2}{2m}\left(3\pi^2\right)^{2/3}n^{5/3}$$

Hence

$$\frac{1}{n}\boldsymbol{\nabla}P = \frac{2}{3}\frac{\hbar^2}{2m}\left(3\pi^2\right)^{2/3}\frac{1}{n^{1/3}}\boldsymbol{\nabla}\delta n \qquad ; \ \delta n \equiv n - n_0$$

Thus from part (a)

$$\frac{2}{3}\frac{\hbar^2}{2m}\left(3\pi^2\right)^{2/3}\frac{1}{n^{1/3}}\boldsymbol{\nabla}\delta n = e_p\boldsymbol{\nabla}\Phi$$

The first of Eqs. (7.122) relates the equilibrium electron Fermi wavenumber to the density

$$k_{\rm F} = (3\pi^2 n_0)^{1/3}$$

Since the l.h.s. is already linear in small quantities, the previous relation can then be written as[40]

$$\frac{2}{3}\frac{\hbar^2k_{\rm F}^2}{2m}\frac{1}{n_0}\boldsymbol{\nabla}\delta n = e_p\boldsymbol{\nabla}\Phi$$

[39]Use $g_s = 2$.

[40]We assume here that $\delta n/n_0 \ll 1$. With the use of our final result $\delta n = Zq_{\rm TF}^2e^{-r\,q_{\rm TF}}/4\pi r$, this condition becomes $\delta n/n_0 = (3/2)(Ze^2/r\varepsilon_{\rm F})e^{-r\,q_{\rm TF}} \ll 1$, which always holds for $Z \to 0$, as well as for large-enough $r\varepsilon_{\rm F}$.

(d) Take the divergence of the above relation, and insert the result from part (b),

$$\frac{2}{3}\frac{\hbar^2 k_F^2}{2m}\frac{1}{n_0}\nabla^2 \delta n = e_p \nabla^2 \Phi$$
$$= -4\pi e_p^2 \left[Z\delta^{(3)}(\mathbf{x}) - \delta n \right]$$

This expression is re-written as

$$\left(\boldsymbol{\nabla}^2 - q_{\rm TF}^2\right)\delta n(\mathbf{x}) = -Z q_{\rm TF}^2 \delta^{(3)}(\mathbf{x}) \qquad ; \; q_{\rm TF} \equiv \left(\frac{6\pi e^2 n_0}{\varepsilon_{\rm F}}\right)^{1/2}$$

(e) This is the Yukawa equation, with solution for the electron charge density $\delta\rho = -e_p \delta n$[41]

$$\delta\rho_{\rm TF}(r) = -Z e_p q_{\rm TF}^2 \frac{e^{-q_{\rm TF}\, r}}{4\pi r} \qquad ; \; \delta\rho = -e_p \delta n$$

This is the Thomas-Fermi result for the induced screening of a point charge in an electron gas (here $r = |\mathbf{x}|$)

(f) The integral over all space gives

$$\int d^3r\, \frac{e^{-q_{\rm TF}\, r}}{4\pi r} = \frac{1}{q_{\rm TF}^2}\int_0^\infty x e^{-x}\, dx = \frac{1}{q_{\rm TF}^2}$$

Therefore one has

$$\int d^3r\, \delta\rho_{\rm TF}(r) = -Z e_p$$

Hence the integrated induced density completely screens the point charge.

Problem 7.25 The Thomas-Fermi theory of the structure of an isolated atom follows from the expressions in Prob. 7.24(b,c), written with

[41]For a spherically symmetric solution, use $\nabla^2 \doteq (1/r)(\partial^2/\partial r^2) r$. The singularity is handled exactly as with the Coulomb Green's function. Integrate over a tiny sphere of radius R around the origin and use Gauss's theorem

$$\int_R d^3r\, \nabla^2 \frac{1}{4\pi r} = \int_S d\mathbf{S}\cdot\boldsymbol{\nabla}\frac{1}{4\pi r} = -R^2 \frac{1}{R^2} = -1$$

$n_0 = 0$ and away from the origin, which will be included through a boundary condition

$$\frac{2}{3}\frac{\hbar^2}{2m}(3\pi^2)^{2/3}\frac{1}{n^{1/3}}\boldsymbol{\nabla} n = e_p \boldsymbol{\nabla}\Phi$$
$$\boldsymbol{\nabla}^2\Phi = 4\pi e_p n$$

(a) Show from the first equation that for a neutral atom the electron density and electrostatic potential are related by

$$e_p\Phi(r) = \frac{\hbar^2}{2m}[3\pi^2 n(r)]^{2/3}$$

(b) Substitute this in the second relation to arrive at

$$\frac{1}{r}\frac{\partial^2}{\partial r^2}(r\Phi) = \kappa\Phi^{3/2}$$

where (recall that a_0 is the Bohr radius)

$$\kappa \equiv \frac{4e_p}{3\pi}\left(\frac{2me_p}{\hbar^2}\right)^{3/2}$$
$$= \frac{8\sqrt{2}}{3\pi}\frac{1}{a_0^2}\left(\frac{1}{e_p/a_0}\right)^{1/2} \qquad ; \; a_0 \equiv \frac{\hbar^2}{m_e e^2}$$

(c) Go to dimensionless variables

$$\phi \equiv \frac{\Phi}{e_p/a_0} \qquad ; \; \rho \equiv \frac{r}{a_0}$$

and show that

$$\frac{1}{\rho}\frac{\partial^2}{\partial\rho^2}(\rho\phi) = \left(\frac{\phi}{b}\right)^{3/2} \qquad ; \; b \equiv \frac{1}{2}\left(\frac{3\pi}{4}\right)^{2/3}$$

(d) Introduce

$$\phi \equiv \frac{Z}{\rho}\chi(x) \qquad ; \; \rho \equiv \frac{b}{Z^{1/3}}x$$

to arrive at the following non-linear Thomas-Fermi differential equation for the shielded electrostatic potential in the atom[42]

$$\sqrt{x}\frac{d^2}{dx^2}\chi(x) = [\chi(x)]^{3/2}$$

(e) Show that the boundary conditions $\Phi \to Ze_p/r$ as $r \to 0$, and $r\Phi \to 0$ as $r \to 0$ for the neutral atom, become

$$\chi(0) = 1 \qquad ; \; \chi(\infty) = 0$$

Solution to Problem 7.25

(a) With only an electron density n, and away from the origin, Probs. 7.24(b,c) give for the Thomas-Fermi theory of an isolated atom

$$\frac{2}{3}\frac{\hbar^2}{2m}(3\pi^2)^{2/3}\frac{1}{n^{1/3}}\boldsymbol{\nabla}n = e_p\boldsymbol{\nabla}\Phi$$
$$\boldsymbol{\nabla}^2\Phi = 4\pi e_p n$$

The first equation is re-written as

$$\frac{\hbar^2}{2m}(3\pi^2)^{2/3}\boldsymbol{\nabla}n^{2/3} = e_p\boldsymbol{\nabla}\Phi$$

This equation is immediately integrated to give

$$e_p\Phi(r) = \frac{\hbar^2}{2m}[3\pi^2 n(r)]^{2/3}$$

where we have incorporated the boundary condition that for a neutral atom, both $n(r)$ and $\Phi(r)$ vanish as $r \to \infty$.

(b) Substitute this relation for $n(r)$ in the second equation in part (a). Then for a spherically symmetric $\Phi(r)$

$$\frac{1}{r}\frac{\partial^2}{\partial r^2}(r\Phi) = \kappa\Phi^{3/2}$$

[42]The shielded electrostatic potential in the atom is here given by

$$\Phi = \frac{Ze_p}{r}\chi(x) \qquad ; \; x = \frac{Z^{1/3}}{ba_0}r$$

where

$$\kappa = \frac{4e_p}{3\pi}\left(\frac{2me_p}{\hbar^2}\right)^{3/2}$$
$$= \frac{8\sqrt{2}}{3\pi}\frac{1}{a_0^2}\left(\frac{1}{e_p/a_0}\right)^{1/2} \qquad ; \; a_0 \equiv \frac{\hbar^2}{m_e e^2}$$

Here the last relation explicitly exhibits the dimensions in terms of fundamental constants.

(c) We can now go over to dimensionless variables[43]

$$\phi \equiv \frac{\Phi}{e_p/a_0} \qquad ; \; \rho \equiv \frac{r}{a_0}$$

It follows directly that

$$\frac{1}{\rho}\frac{\partial^2}{\partial\rho^2}(\rho\phi) = \left(\frac{\phi}{b}\right)^{3/2} \qquad ; \; b \equiv \frac{1}{2}\left(\frac{3\pi}{4}\right)^{2/3}$$

(d) With the introduction of

$$\phi \equiv \frac{Z}{\rho}\chi(x) \qquad ; \; \rho \equiv \frac{b}{Z^{1/3}}x$$

we have

$$\frac{Z}{b^3}\frac{1}{x}\frac{d^2}{dx^2}[Z\chi(x)] = \left[\frac{Z^{4/3}\chi(x)}{bx}\right]^{3/2}\frac{1}{b^{3/2}}$$

Hence, we arrive at the following non-linear Thomas-Fermi differential equation for the shielded electrostatic potential in the atom

$$\sqrt{x}\frac{d^2}{dx^2}\chi(x) = [\chi(x)]^{3/2}$$

(e) The electrostatic potential at the origin arises from the point nuclear charge, and in the neutral atom, the potential must vanish faster than $1/r$ as $r \to \infty$. The boundary conditions $\Phi \to Ze_p/r$ as $r \to 0$, and $r\Phi \to 0$ as $r \to 0$ for the neutral atom then become

$$\chi(0) = 1 \qquad ; \; \chi(\infty) = 0$$

[43] Always a good practice!

Problem 7.26 Integrate the non-linear differential equation in Prob. 7.25(d,e) numerically, starting at (or very near) the origin and changing the intitial slope $\chi'(0)$ until the solution vanishes for large x.[44]

This calculation represents one of the first published applications of modern computers to physics [Feynman, Metropolis, and Teller (1949)].

Solution to Problem 7.26

The Thomas-Fermi equation in Prob. 7.25(d,e) can be integrated numerically by starting at the origin where $\chi(0) = 1$ and choosing a slope $\chi'(0)$. One then just steps out in x, and re-adjusts $\chi'(0)$ until a solution is found that decreases to zero for large x. Numerical results obtained using Mathcad 11 and the Runge-Kutta algorithm are shown in Fig. 7.8. It is easiest to start at a very small, but finite x. The calculation shown in Fig. 7.8 actually starts from $[\chi(x_0), \chi'(x_0)] = [1 + x_0\chi'(x_0), -1.568\cdots]$ with $x_0 = 1 \times 10^{-4}$.

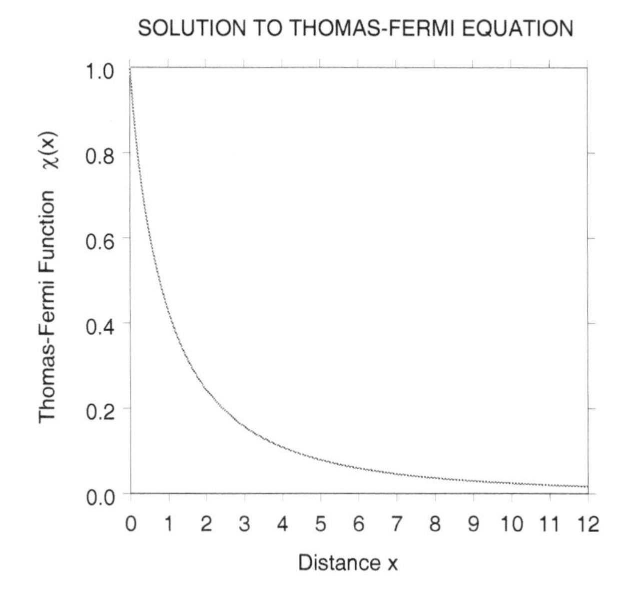

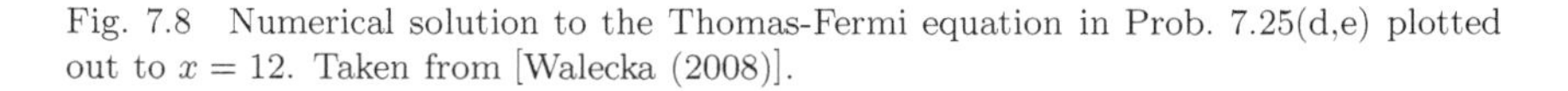

Fig. 7.8 Numerical solution to the Thomas-Fermi equation in Prob. 7.25(d,e) plotted out to $x = 12$. Taken from [Walecka (2008)].

[44] See Fig. 5.11 in [Walecka (2008)].

Problem 7.27 Assume the nuclear interactions are equivalent to a slowly varying potential $-U(r)$. Within any small volume element, assume the nucleons form a non-interacting Fermi gas filled up to an energy $-\mathcal{E}$.

(a) Show that in equilibrium $\mathcal{E}$ must be constant throughout the nucleus;

(b) Derive the Thomas-Fermi result for the baryon number density[45]

$$n(r) = \frac{2}{3\pi^2}\left(\frac{2m}{\hbar^2}\right)^{3/2}[U(r)-\mathcal{E}]^{3/2}$$

(c) Derive this result by balancing the hydrostatic force $-\boldsymbol{\nabla}P$ and the force from the potential $n\boldsymbol{\nabla}U$;

(d) Show the total baryon number and energy are given by

$$B = \int_0^R d^3r\, n(r) \qquad ; \; U(R) = \mathcal{E}$$

$$E = \int_0^R d^3r \left[\frac{3}{5}\frac{\hbar^2}{2m}\left(\frac{3\pi^2}{2}\right)^{2/3} n(r)^{5/3} - U(r)n(r)\right]$$

(e) Construct the zero-temperature thermodynamic potential $\Phi = E - \mu B$, where μ is the chemical potential. Show that if Φ is made stationary under arbitrary variations of the density $\delta n(r)$, one recovers the result in part (b). Identify μ.

This is the simplest example of *density functional theory*, a now widely-used tool for calculating the structure of many-body assemblies [Kohn (1999)].

Solution to Problem 7.27

(a) In the Thomas-Fermi approximation, the energy levels are filled locally in the potential $-U(r)$ up to a Fermi level $\varepsilon_{\rm F}(r)$ at the overall energy $-\mathcal{E}$ [see Fig. 7.9]. The ground state will correspond to that configuration with minimum energy, and $\mathcal{E}$ must then be constant throughout the nucleus. If it were not the case, one could lower the energy of the assembly by redistributing some of the baryon density $n(r)$ from regions of higher to lower energy.

(b) In nuclear matter with $(n\uparrow, n\downarrow, p\uparrow, p\downarrow)$, the baryon number density is (see Prob. 7.11)

$$n(r) = \frac{2k_{\rm F}^3}{3\pi^2}$$

[45] Assume equal numbers of neutrons and protons (recall Prob. 7.11).

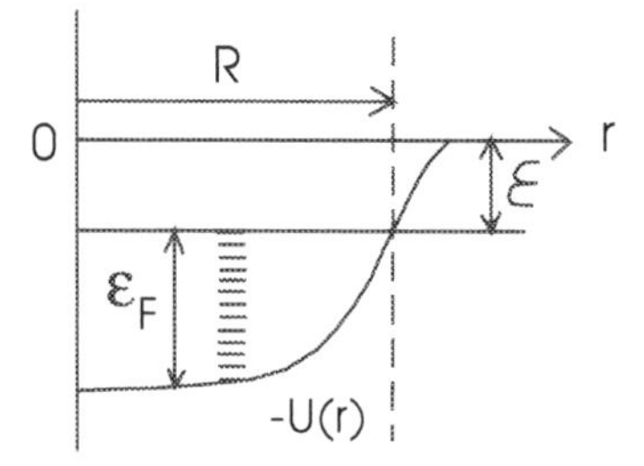

Fig. 7.9 Nuclear interactions equivalent to a slowly varying potential $-U(r)$. Within any small volume element, nucleons form a non-interacting Fermi gas, with a local density $n(r)$, filled up to an energy $-\mathcal{E}$.

The Fermi energy is then

$$\varepsilon_{\rm F}(r) = \frac{\hbar^2 k_{\rm F}^2(r)}{2m_p} = \frac{\hbar^2}{2m_p}\left[\frac{3\pi^2 n(r)}{2}\right]^{2/3}$$

(c) The energy relation from part (a) is [see Fig. 7.9]

$$\varepsilon_{\rm F}(r) - U(r) = -\mathcal{E}$$

Substitution of the expression for $\varepsilon_{\rm F}(r)$ in part (b), and solution for $n(r)$, then yields the Thomas-Fermi result for the baryon number density

$$n(r) = \frac{2}{3\pi^2}\left(\frac{2m_p}{\hbar^2}\right)^{3/2}[U(r) - \mathcal{E}]^{3/2}$$

(d) Consider a small volume with sides of area $\mathcal{A}$ perpendicular to the radial direction, thickness dr, and volume $\mathcal{A}\,dr = v$. The outward pressure force on this small volume is

$$F_P = P(r)\mathcal{A} - P(r+dr)\mathcal{A} = -\frac{dP}{dr}\mathcal{A}\,dr = -v\frac{dP}{dr}$$

The inward nuclear attraction on the nv baryons in this volume v is

$$F_N = nv\frac{d[-U(r)]}{dr}$$

If these forces are balanced, then

$$\frac{dP(r)}{dr} = n\frac{dU(r)}{dr}$$

The pressure exerted by this Fermi gas is given in Eqs. (7.122)

$$P = \frac{2}{5}\frac{\hbar^2}{2m_p}\left(\frac{3\pi^2}{2}\right)^{2/3} n^{5/3}$$
$$\frac{dP}{dr} = \frac{2}{3}\frac{\hbar^2}{2m_p}\left(\frac{3\pi^2}{2}\right)^{2/3} n^{2/3}\frac{dn}{dr}$$

The balance of forces then yields

$$\frac{dU(r)}{dr} = \frac{\hbar^2}{2m_p}\left(\frac{3\pi^2}{2}\right)^{2/3} \frac{d}{dr}[n(r)]^{2/3}$$

Integration of this relation reproduces our previous result

$$n(r) = \frac{2}{3\pi^2}\left(\frac{2m_p}{\hbar^2}\right)^{3/2} [U(r) - \mathcal{E}]^{3/2}$$

where $\mathcal{E}$ is now simply the constant of integration.

(d) We have $n(r)\,d^3r$ baryons in the volume d^3r. The average kinetic energy per baryon is $3\varepsilon_{\rm F}(r)/5$ and the potential energy per baryon is $-U(r)$. Hence the total number of baryons and energy of the assembly are

$$B = \int_0^R d^3r\, n(r) \qquad ; \; U(R) = \mathcal{E}$$
$$E = \int_0^R d^3r \left[\frac{3}{5}\frac{\hbar^2}{2m}\left(\frac{3\pi^2}{2}\right)^{2/3} n(r)^{5/3} - U(r)n(r)\right]$$

Here it has been observed that the Thomas-Fermi density goes to zero at a radius R, where $U(R) = \mathcal{E}$ [see Fig. 7.9].

(e) The thermodynamic potential at zero temperature and zero pressure is constructed as in Eqs. (6.18)[46]

$$\tilde{\Phi} = E + PV - TS - \mu B$$
$$= E - \mu B \qquad ; \; (T, P) = 0$$

where μ is the chemical potential. Thus

$$\tilde{\Phi} = \int_0^R d^3r \left[\frac{3}{5}\frac{\hbar^2}{2m}\left(\frac{3\pi^2}{2}\right)^{2/3} n(r)^{5/3} - U(r)n(r) - \mu n(r)\right]$$

[46] To arrive at the stated equilibrium condition, we formally include a Legendre transformation to the enthalpy in $\tilde{\Phi}$, as in Probs. 6.4 and A.4; however, since $P = 0$ this does not matter, and $\tilde{\Phi} \equiv \Phi$.

The condition for equilibrium in Eq. (6.22) becomes

$$\delta\tilde{\Phi}\Big|_{\mu} \geq 0 \qquad\qquad ; (T,P) = 0$$

We set $\delta\tilde{\Phi} = 0$ at fixed μ to locate the minimum

$$\delta\tilde{\Phi} = \int_0^R d^3r \left[\frac{\hbar^2}{2m}\left(\frac{3\pi^2}{2}\right)^{2/3} n(r)^{2/3} - U(r) - \mu\right]\delta n(r) = 0$$

With arbitrary variations in the density $\delta n(r)$, this gives

$$\varepsilon_{\rm F}(r) - U(r) = \mu$$

Identification of the chemical potential as [see Fig. 7.9]

$$\mu = -\mathcal{E}$$

reproduces the previous analysis.

As stated in the problem, this is the simplest example of *density functional theory*, a now widely-used tool for calculating the structure of many-body assemblies [Kohn (1999)].

Problem 7.28 The quanta of the sound-wave excitations in a solid can be considered to form a Bose gas of *phonons*, with a phonon energy of $\varepsilon = h\nu$, frequency cut-off of ν_m, and spectral density given by Eq. (5.29).

(a) Show that the phonon chemical potential must vanish [recall Eq. (7.30)];

(b) Combine the resulting Bose-Einstein distribution function in the second of Eqs. (7.18) with the phonon energy to re-derive the Debye expressions for the energy and heat capacity of the phonon gas in Eqs. (5.33).

Solution to Problem 7.28

(a) The number of phonons is not conserved. It then follows exactly as in Eqs. (7.29)–(7.35) that the phonon chemical potential $\mu_{\rm ph}$ must vanish[47]

$$\mu_{\rm ph} = 0 \qquad\qquad ; \lambda_{\rm ph} = e^{\mu_{\rm ph}/k_{\rm B}T} = 1$$

(b) The Bose-Einstein distribution function in the second of Eqs. (7.18) then gives the number of phonons in each mode ν at the temperature T

$$n_\nu = \frac{1}{e^{h\nu/k_{\rm B}T} - 1}$$

[47]Compare Prob. 8.29.

The phonon spectral distribution is given in Eq. (5.29)

$$g(\nu)d\nu = 9N\frac{\nu^2 d\nu}{\nu_m^3} \qquad ; \ \nu \leq \nu_m$$

The energy per phonon is $\varepsilon = h\nu$. These expressions can now be combined to give the total phonon energy

$$E = 9N\int_0^{\nu_m} \frac{\nu^2 d\nu}{\nu_m^3}\frac{h\nu}{e^{h\nu/k_\text{B}T}-1}$$

With the introduction of the Debye temperature in Eq. (5.31)

$$\theta_D \equiv \frac{h\nu_m}{k_\text{B}}$$

the energy is re-written as

$$E = 9Nk_\text{B}T\left(\frac{T}{\theta_D}\right)^3\int_0^{\theta_D/T}\frac{u^3 du}{e^u - 1} \qquad ; \ u = \frac{h\nu}{k_\text{B}T}$$

The heat capacity then follows just as in the second of Eqs. (5.33)

$$C_V = 9Nk_\text{B}\left(\frac{T}{\theta_D}\right)^3\int_0^{\theta_D/T}\frac{u^4 e^u \, du}{(e^u-1)^2}$$

These are the Debye expressions for the temperature-dependent energy and heat capacity of the phonon gas in Eqs. (5.33).[48]

[48]Here E omits the additional temperature-independent zero-point energy of the oscillators [see Eq. (5.4)].

Chapter 8

Special Topics

Problem 8.1 Show that as long as the volume V is held fixed, one can just as well use the full Helmholtz free energy $A_c \to A_c + A_g$ in computing the free energy difference in a perfect solution at a given (M_1, M_2) in Eq. (8.17).

Solution to Problem 8.1

Suppose one has (N_A, N_B) atoms of components (A, B), with (N_1, N_2) of each type in the gas phase and (M_1, M_2) of each type in the condensed phase. The Helmholtz free energy of the perfect solution is then given in Eqs. (8.6)

$$
\begin{aligned}
A &= -k_B T \ln (\text{P.F.}) = A_c + A_g \\
A_c &= -k_B T \ln \left\{ \frac{(M_1 + M_2)!}{M_1! M_2!} \left[(\text{p.f.})_A^c\right]^{M_1} \left[(\text{p.f.})_B^c\right]^{M_2} \right\} \\
A_g &= -k_B T \ln \left\{ \frac{\left[(\text{p.f.})_A^g\right]^{N_1}}{N_1!} \frac{\left[(\text{p.f.})_B^g\right]^{N_2}}{N_2!} \right\}
\end{aligned}
$$

Here, for ease of writing, we again suppress the star on the mean numbers (M_1, N_1, M_2, N_2). From Eqs. (8.9), the numbers (N_1, N_2) in the gas are given in terms of the numbers (M_1, M_2) in the condensed phase by

$$
\begin{aligned}
N_1 &= \frac{M_1}{M_1 + M_2} \frac{(\text{p.f.})_A^g}{(\text{p.f.})_A^c} \\
N_2 &= \frac{M_2}{M_1 + M_2} \frac{(\text{p.f.})_B^g}{(\text{p.f.})_B^c}
\end{aligned}
$$

Define the Helmholtz free energy of the pure materials (A, B) by [compare Eqs. (8.16)]

$$A_1 = -k_B T \ln \left[(\text{p.f.})^c_A\right]^{M_1} - k_B T \ln \frac{\left[(\text{p.f.})^g_A\right]^{N_1}}{N_1!} \qquad ; \text{ pure (A, B)}$$

$$A_2 = -k_B T \ln \left[(\text{p.f.})^c_B\right]^{M_2} - k_B T \ln \frac{\left[(\text{p.f.})^g_B\right]^{N_2}}{N_2!}$$

Now take the difference between the full Helmholtz free energy $A = A_c + A_g$ and the pure (A, B) values $A_1 + A_2$, at fixed (M_1, M_2, V)

$$\Delta A = A_c + A_g - A_1 - A_2 = -k_B T \ln \frac{(M_1 + M_2)!}{M_1! M_2!}$$

This reproduces the free energy difference in a perfect solution at a given (M_1, M_2) in Eq. (8.17). There is no additional contribution to the free-energy difference coming from mixing in the gas phase.

Problem 8.2 Show the employment of Eq. (8.26) in Eq. (8.24) gives Raoult's law.

Solution to Problem 8.2

Suppose one has (N_A, N_B) atoms of components (A, B), with (N_1, N_2) of each type in the gas phase and (M_1, M_2) of each type in the condensed phase, which we now treat as a *regular solution*, utilizing Eqs. (8.24) and (8.25). The Helmholtz free energy is again given by Eqs. (8.6)[1]

$$A = -k_B T \ln (\text{P.F.}) = A_c + A_g$$

$$A_c = -k_B T \ln \left\{ \frac{(M_1 + M_2)!}{M_1! M_2!} \left[(\text{p.f.})^c_A\right]^{M_1} \left[(\text{p.f.})^c_B\right]^{M_2} \right\}$$

$$A_g = -k_B T \ln \left\{ \frac{\left[(\text{p.f.})^g_A\right]^{N_1}}{N_1!} \frac{\left[(\text{p.f.})^g_B\right]^{N_2}}{N_2!} \right\}$$

The only change is that the partition functions for the condensed phase are now given by

$$(\text{p.f.})^c_A = f_A(T, v) \qquad ; \ (\text{p.f.})^c_B = f_B(T, v)$$

The analysis then proceeds exactly as in Eqs. (8.4)–(8.15), with the result that the partial pressures satisfy Eq. (8.15)

$$p = x_1 p_1^0(T) + (1 - x_1) p_2^0(T) \qquad ; \text{ Raoult's law}$$

[1]As above, we suppress the star on the mean numbers (M_1, N_1, M_2, N_2).

This is Raoult's law.

Problem 8.3 Make a good numerical plot of $\Delta\tilde{A} = \Delta A/k_{\rm B}T(N_A+N_B)$ in Fig. 8.5 in the text for various values of the parameter $\gamma \equiv \omega Z/k_{\rm B}T$ in the first of Eqs. (8.44). Verify that the phase transition occurs at $\gamma = 2$.

Solution to Problem 8.3

The Helmholtz free energy of mixing for a regular solution in Eq. (8.44), with an unlike nearest-neighbor pair creation energy of ω [see Eq. (8.32)] treated in the Bragg-Williams approximation, is shown in Fig. 8.1 for three values of $\gamma \equiv \omega Z/k_{\rm B}T$ (compare Fig. 8.5 in the text). The phase transition occurs at $\gamma = 2$.

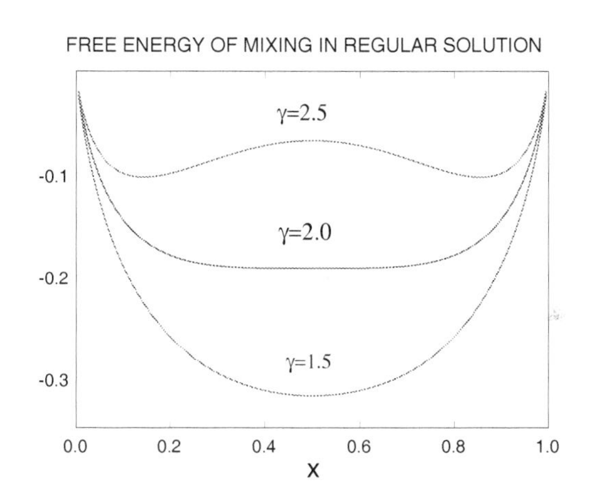

Fig. 8.1 Numerical plot of $\Delta\tilde{A} = \Delta A/k_{\rm B}T(N_A + N_B)$ for a regular solution with interactions in the Bragg-Williams approximation [see Eqs. (8.44)]. Here $\gamma \equiv \omega Z/k_{\rm B}T$. There are three curves: one above T_C at $\gamma = 1.5$; one below T_C at $\gamma = 2.5$; and one at T_C with $\gamma = 2$. (Compare Fig. 8.5 in the text.)

Problem 8.4 Suppose one starts with (N_A, N_B) systems in the condensed phase below T_C, with a mole fraction x_0 lying between the two crosses in Fig. 8.5 in the text. The mole fractions of the two new phases are $x = 1/2 \pm \delta$.

(a) How much of each new phase will be formed?

(b) Show that the new phases indeed have a lower free energy than the original solution.

Solution to Problem 8.4

(a) The original mole fractions are

$$x_0 = \frac{N_A}{N_A + N_B} \qquad ; \; 1 - x_0 = \frac{N_B}{N_A + N_B}$$

Denote the phase at the right cross in Fig. 8.5 in the text by phase 2, and that at the left cross by phase 1. The mole fractions in phase 1 then satisfy

$$\frac{N_{A1}}{N_{A1} + N_{B1}} = \frac{1}{2} - \delta \qquad ; \; \frac{N_{B1}}{N_{A1} + N_{B1}} = 1 - \frac{N_{A1}}{N_{A1} + N_{B1}} = \frac{1}{2} + \delta$$

In phase 2, the mole fractions are

$$\frac{N_{A2}}{N_{A2} + N_{B2}} = \frac{1}{2} + \delta \qquad ; \; \frac{N_{B2}}{N_{A2} + N_{B2}} = 1 - \frac{N_{A2}}{N_{A2} + N_{B2}} = \frac{1}{2} - \delta$$

It follows that

$$N_{A1}\left(\frac{1}{2} + \delta\right) - N_{B1}\left(\frac{1}{2} - \delta\right) = 0$$

$$(N_A - N_{A1})\left(\frac{1}{2} - \delta\right) - (N_B - N_{B1})\left(\frac{1}{2} + \delta\right) = 0$$

where we have used

$$N_A = N_{A1} + N_{A2} \qquad ; \; N_B = N_{B1} + N_{B2}$$

Addition of the above two simultaneous equations, and division by $N_A + N_B$, gives

$$2\delta\left[\frac{N_{A1} + N_{B1}}{N_A + N_B}\right] = \left(\frac{1}{2} + \delta\right)\frac{N_B}{N_A + N_B} - \left(\frac{1}{2} - \delta\right)\frac{N_A}{N_A + N_B}$$

Call r the fraction of the number of systems that ends up in phase 1

$$r = \frac{N_{A1} + N_{B1}}{N_A + N_B} \qquad ; \; 1 - r = \frac{N_{A2} + N_{B2}}{N_A + N_B}$$

Then the above reads

$$2r\delta = \left(\frac{1}{2} + \delta\right) - x_0$$

$$2(1 - r)\delta = x_0 - \left(\frac{1}{2} - \delta\right)$$

Therefore, the ratio of the total number of systems in each phase is

$$\frac{r}{1-r} = \frac{N_{A1} + N_{B1}}{N_{A2} + N_{B2}} = \frac{(1/2 + \delta) - x_0}{x_0 - (1/2 - \delta)}$$

This is simply the ratio of the distance of the right cross from x_0, to the distance of x_0 from the left cross along the horizontal dotted straight line in Fig. 8.5 in the text.

(b) Let $\Delta\tilde{A}_2$ be the Helmholtz free energy of mixing per system at the right cross and $\Delta\tilde{A}_1$ that at the left cross. Then

$$\Delta\tilde{A}_1 = \Delta\tilde{A}_2 \equiv \Delta\tilde{A}_{\text{min}}$$

The free energy of mixing of the phase-separated assembly follows from part (a) as

$$\Delta A_{\text{p-s}} = (N_{A1} + N_{B1})\Delta\tilde{A}_1 + (N_{A2} + N_{B2})\Delta\tilde{A}_2 = (N_A + N_B)\Delta\tilde{A}_{\text{min}}$$

The free energy of mixing of the homogeneous assembly at the mole fraction x_0 is given by the solid curve labeled $\Delta\tilde{A}$ in Fig. 8.5 in the text

$$\Delta A_{\text{hom}} = (N_A + N_B)\Delta\tilde{A}$$

Clearly

$$\Delta A_{\text{p-s}} < \Delta A_{\text{hom}} \qquad ; \frac{1}{2} - \delta \leq x_0 \leq \frac{1}{2} + \delta$$

The phase-separated assembly has a lower Helmholtz free energy than that of the homogeneous assembly for all x_0 lying between the two crosses in Fig. 8.5 in the text.

Problem 8.5 Consider a two-dimensional square lattice of N_s sites. Suppose N_p particles are placed on those sites. Assume there is a constant energy shift of $-\epsilon$ when two particles sit next to each other on the lattice (see Fig. 8.2).

(a) Periodic boundary conditions will be assumed in both directions so that all sites are equivalent.[2] Show that a physical realization of the periodic boundary conditions is achieved by putting the lattice on the surface of a torus. Explain the relevance to Fig. 8.2;

[2]One can alternatively assume a very large lattice, so that boundary effects are unimportant.

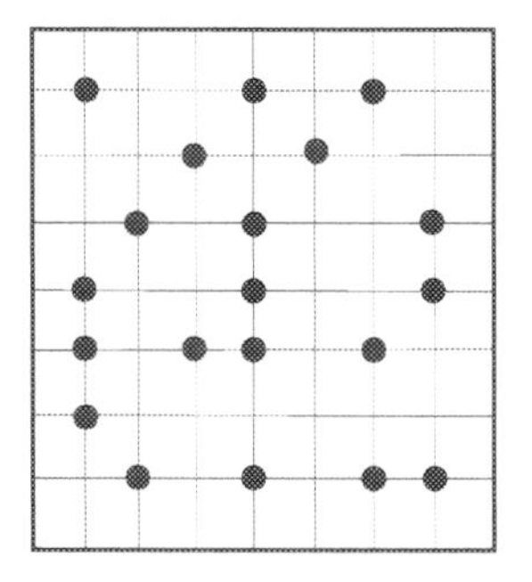

Fig. 8.2 Lattice gas. There are N_p particles on N_s sites, with an energy shift of $-\epsilon$ for each nearest-neighbor pair.

(b) Work in the canonical ensemble. Show the configuration free energy and configuration partition function for this *lattice gas* are given by

$$\begin{aligned} A_C &= -k_{\rm B}T \ln{(\text{P.F.})_C} \\ (\text{P.F.})_C &= \sum_{N_{pp}} g(N_s, N_p, N_{pp}) e^{\epsilon N_{pp}/k_{\rm B}T} \end{aligned}$$

Define all quantities.

Solution to Problem 8.5

(a) Periodic boundary conditions in one dimension are achieved by putting the assembly around a cylinder. When you go around the cylinder, you come back where you started. Periodic boundary conditions in the second dimension can then be physically achieved by joining the ends of the cylinder and converting the cylinder into a *torus* (that is, a *doughnut*). Now when you go around in the second dimension, you again come back to where you started.[3]

The relevance to Fig. 8.2 is that with periodic boundary conditions, the sites on the r.h.s. are, in fact, *identical* to the sites on the l.h.s.; similarly, the sites at the top are identical to those at the bottom.

(b) The Einstein model of independent localized systems is discussed in section 8.1.2. With N_p particles on N_p sites, the canonical partition function is given by Eq. (8.23)

$$(\text{P.F.}) = [f(T, v)]^{N_p}$$

[3]Compare Prob. 8.13.

If the N_p particles are distributed on N_s equivalent sites, the partition function is [see Eq. (8.24)]

$$(\text{P.F.}) = [f(T,v)]^{N_p}\, g(N_p, N_s)$$

where $g(N_p, N_s)$ *is the number of complexions of* N_p *particles on* N_s *sites*

$$g(N_p, N_s) = \frac{N_s!}{N_p!(N_s - N_p)!}$$

Now suppose there is an additional interaction energy $-\varepsilon$ for two nearest neighbors. Then

$$g(N_p, N_s) \to \sum_{N_{pp}} g(N_p, N_s, N_{pp}) e^{\varepsilon N_{pp}/k_{\text{B}}T}$$

where $g(N_p, N_s, N_{pp})$ *is the number of complexions of given* (N_p, N_s, N_{pp}), *where* N_{pp} *is the number of nearest neighbors in the complexion.* The Helmholtz free energy is then given by

$$\begin{aligned} A &= -k_{\text{B}}T \ln (\text{P.F.}) \\ &= A_p + A_C \end{aligned}$$

where A_p is the free energy of the N_p localized non-interacting systems,[4] and the *configurational free energy* is

$$\begin{aligned} A_C &= -k_{\text{B}}T \ln (\text{P.F.})_{\text{C}} \\ (\text{P.F.})_{\text{C}} &= \sum_{N_{pp}} g(N_p, N_s, N_{pp}) e^{\varepsilon N_{pp}/k_{\text{B}}T} \end{aligned}$$

Here $(\text{P.F.})_{\text{C}}$ is the *configuration partition function.*

Problem 8.6 (a) The Bragg-Williams approximation for the lattice gas problem formulated in Prob. 8.5 replaces N_{pp} in the sum in the configuration partition function by the value for a random distribution of particles on the lattice. Show this number is given by

$$\overset{o}{N}_{pp} = \frac{2N_p^2}{N_s}$$

[4] From above, $A_p = -k_{\text{B}}T \ln [f(T,v)]^{N_p}$.

(b) Show that if $(N_s, N_p) \gg 1$, the configuration free energy is then given by

$$\frac{A_C}{k_B T N_s} = -\frac{2\epsilon}{k_B T}x^2 + [x \ln x + (1-x)\ln(1-x)]$$
$$x \equiv \frac{N_p}{N_s}$$

Solution to Problem 8.6

(a) The *Bragg-Williams approximation* uses the random number of nearest neighbor pairs in calculating the configuration partition function. Start at one corner and move through the N_s sites on the the lattice. The probability that you will find a particle on this site is N_p/N_s. Check on the neighbors in the positive (x, y) coordinate directions, there are two of them, giving a total number of nearest neighbor possibilites on the lattice of $2N_s$. The random probability that you will find a particle on the neighboring site is again N_p/N_s. By the time you have gone through the whole lattice, you will have examined all possible nearest-neighbor pairs.[5] Hence

$$\overset{o}{N}_{pp} = 2N_s \times \frac{N_p}{N_s} \times \frac{N_p}{N_s} = 2\frac{N_p^2}{N_s} \qquad ; \text{Bragg-Williams}$$

(b) The Bragg-Williams approximation then makes the following replacement in the configuration partition function

$$e^{\varepsilon N_{pp}/k_B T} \rightarrow e^{\varepsilon \overset{o}{N}_{pp}/k_B T}$$

which removes this weighting from the sum. In the remaining sum of $g(N_s, N_p, N_{pp})$ over all N_{pp}, the nearest neighbor sorting of the complexions has now been removed from the counting, and we simply recover the total number of complexions $g(N_s, N_p)$

$$\sum_{N_{pp}} g(N_p, N_s, N_{pp}) = g(N_s, N_p) = \frac{N_s!}{N_p!(N_s - N_p)!}$$

Hence

$$(\text{P.F.})_C \doteq \frac{N_s!}{N_p!(N_s - N_p)!} \exp\left\{\frac{2\varepsilon N_p^2}{N_s k_B T}\right\} \qquad ; \text{Bragg-Williams}$$

[5] Compare the argument in Prob. 8.15.

With large (N_s, N_p), we can use Stirling's approximation $\ln N! \approx N \ln N - N$, and this gives

$$
\begin{aligned}
-\frac{A_C}{k_{\rm B}T} &= N_s \ln N_s - N_s - N_p \ln N_p + N_p - (N_s - N_p)\ln(N_s - N_p) \\
&\quad + (N_s - N_p) + \frac{2\varepsilon N_p^2}{N_s k_{\rm B}T} \\
&= N_s \ln \frac{N_s}{N_s - N_p} - N_p \ln \frac{N_p}{N_s - N_p} + \frac{2\varepsilon N_p^2}{N_s k_{\rm B}T} \qquad ; \ (N_s, N_p) \gg 1
\end{aligned}
$$

Define the fractional occupancy by

$$
x \equiv \frac{N_p}{N_s} \qquad ; \text{ fractional occupancy}
$$

Then the Bragg-Williams approximation for the configurational free energy per site for the lattice gas is

$$
\frac{A_C}{N_s} = -2\varepsilon x^2 + k_{\rm B}T\left[x \ln x + (1-x)\ln(1-x)\right]
$$

Problem 8.7 (a) At high temperature, the free energy in Prob. 8.6(b) is concave up as a function of the filling fraction x. Show that as soon as such a free energy develops a region that is concave down, then it possible to lower the free energy of the assembly by separating it into two phases with fractions (x_1, x_2) determined from the points of tangency of a straight line tangent to the curve at two points.

(b) How much of each phase will be formed?

(c) At what temperature will this phase separation occur in this simple lattice gas model?

Solution to Problem 8.7

(a) Call a_c the configurational free energy per site in Prob. 8.6(b)

$$
a_c \equiv \frac{A_C}{N_s} = -2\varepsilon x^2 + k_{\rm B}T\left[x \ln x + (1-x)\ln(1-x)\right]
$$

Take two derivatives of this expression

$$
\begin{aligned}
\frac{\partial a_c}{\partial x} &= -4\varepsilon x + k_{\rm B}T\left[\ln x + 1 - \ln(1-x) - 1\right] \\
\frac{\partial^2 a_c}{\partial x^2} &= -4\varepsilon + k_{\rm B}T\left[\frac{1}{x} + \frac{1}{1-x}\right]
\end{aligned}
$$

Evidently

$$\frac{\partial^2 a_c}{\partial x^2} \geq -4\varepsilon + 4k_{\text{B}}T$$

Now observe that

- If $k_{\text{B}}T > \varepsilon$, then $\partial^2 a_c/\partial x^2 > 0$, and the curve is *always concave up*;
- If $k_{\text{B}}T < \varepsilon$, then the second derivative $\partial^2 a_c/\partial x^2$ goes $(+, -, +)$ as x goes from 0 to 1, and there is a region in x where the curve is *concave down* [see Fig. 8.3]. The transition temperature for this to occur is

$$k_{\text{B}}T_C = \varepsilon$$

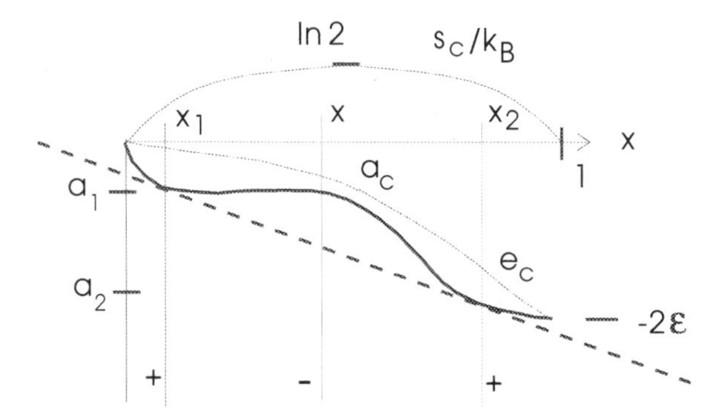

Fig. 8.3 Configurational free energy per site a_c for the lattice gas in the Bragg-Williams approximation (solid curve) for the case where there is a region in x where the curve is *concave down*. Here $a_c \equiv e_c - Ts_c$.

(b) Draw a straight line tangent to the curve $a_c(x)$ at the two points $a_1(x_1)$ and $a_2(x_2)$ as indicated in Fig. 8.3. It is verified below that at any intermediate fractional occupancy $x_1 \leq x \leq x_2$, the free energy of the assembly will lie below the curve $a_c(x)$ if it separates into two phases characterized by the points of tangency.

Let αN_s sites have the composition x_2 with a_2, and $(1-\alpha)N_s$ sites have the composition x_1 with a_1. Then the number of particles in each phase will be

$$N_2^p = (\alpha N_s)x_2$$
$$N_1^p = [(1-\alpha)N_s]x_1$$

The total number of particles is

$$N_p = N_1^p + N_2^p = xN_s = [\alpha x_2 + (1-\alpha)x_1]N_s$$

Therefore

$$x = \alpha x_2 + (1-\alpha)x_1$$
$$\alpha = \frac{x - x_1}{x_2 - x_1}$$

The fraction of the lattice α with phase 2 is just the fractional distance the initial composition x lies between x_2 and x_1.

Let us examine the total phase-separated configurational free energy

$$N_s a = (\alpha N_s)a_2 + [(1-\alpha)N_s]a_1$$
$$a = a_1 + \alpha(a_2 - a_1)$$

We observe that

(1) This is a straight line $a(\alpha)$;
(2) The line has slope $a'(\alpha) = (a_2 - a_1)$ and intercept $a(0) = a_1$
(3) This is just the line of tangency in Fig. 8.3, which *obviously lies below the curve* $a_c(x)$ *for* $x_1 \leq x \leq x_2$.

Hence there will be a *phase separation* in this lattice gas at low enough temperature.[6]

(c) The critical temperature was obtained above as

$$k_B T_C = \varepsilon$$

Problem 8.8 Show that for temperatures $T < T_C$, the solution with the non-zero pair $\pm s$ in Fig. 8.9 in the text yields a lower Helmholtz free energy than the solution at the origin.

Solution to Problem 8.8

Consider the difference in configuration Helmholtz free energy for the order-disorder transition in the Bragg-Williams approximation which follows from Eq. (8.67)

[6]If the lattice is upright in a small gravitational field **g**, for example, then the dense phase will physically separate at the bottom of the lattice.

$$
\begin{aligned}
F(s) &\equiv \frac{2}{Nk_{\rm B}T}[A_C(0) - A_C(s^2)] \\
&= -\left[(1+s)\ln(1+s) + (1-s)\ln(1-s)\right] + \frac{Z\omega}{2k_{\rm B}T}s^2
\end{aligned}
$$

Substitute the expression that determines s from Eqs. (8.63)

$$
\ln\frac{1+s}{1-s} = \frac{Z\omega s}{k_{\rm B}T}
$$

Then

$$
\begin{aligned}
F(s) &= -\left[(1+s)\ln(1+s) + (1-s)\ln(1-s)\right] + \frac{s}{2}\ln\frac{1+s}{1-s} \\
&= -\ln(1-s^2) - \frac{s}{2}\ln\frac{1+s}{1-s}
\end{aligned}
$$

We want to show that $F(s)$ is positive when s is determined as in Fig. 8.9 in the text. Note there is a solution with $0 \leq s \leq 1$, when $T \leq T_C$ where

$$
k_{\rm B}T_C = \frac{1}{2}Z\omega \qquad ; \text{ transition temperature}
$$

Consider the two limiting cases:

- When $T \to 0$, then $s \to 1$, and

$$
F(s) \to -\frac{1}{2}\ln(1-s)
$$

  which is positive.
- When $T \to T_C$, then $s \to 0$, and[7]

$$
F(s) \to \frac{s^4}{6} \qquad ; \; F'(s) \to \frac{2s^3}{3}
$$

  which are both positive.

Take one derivative of $F(s)$ in the interval

$$
\begin{aligned}
F'(s) &= \frac{2s}{1-s^2} - \frac{1}{2}\ln\frac{1+s}{1-s} - \frac{s}{2}\left[\frac{1}{1+s} + \frac{1}{1-s}\right] \\
&= \frac{s}{1-s^2} - \frac{1}{2}\ln\frac{1+s}{1-s}
\end{aligned}
$$

[7]Use $\ln(1+x) = x - x^2/2 + x^3/3 + \cdots$.

Now take one more

$$F''(s) = \frac{1}{1-s^2} + \frac{2s^2}{(1-s^2)^2} - \frac{1}{2}\left[\frac{1}{1+s} + \frac{1}{1-s}\right]$$
$$= \frac{2s^2}{(1-s^2)^2}$$

Since $F(s)$ is positive with positive slope for small s, and the curvature is everywhere positive, then $F(s)$ is indeed positive in the interval $0 \leq s \leq 1$.

Problem 8.9 (a) Solve Eq. (8.64) numerically for $s(T/T_C)$, and make a good plot of the molar entropy S_C/R in Fig. 8.10 in the text;

(b) Use the results from (a), compute the heat capacity from Eq. (8.69), and make a good plot of the Bragg-Williams curve in Fig. 8.11 in the text.

Solution to Problem 8.9

(a) The molar configuration entropy S_C/R for an order-disorder transition in the Bragg-Williams approximation calculated with Mathcad 7 from Eqs. (8.68) and (8.64) is shown in Fig. 8.4.

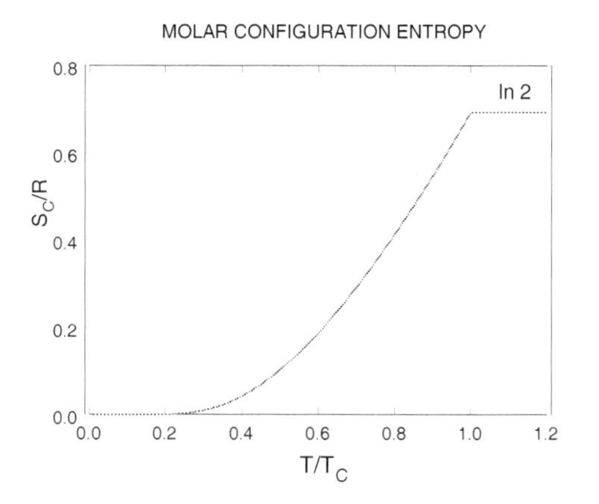

Fig. 8.4 Numerical calculation from Eqs. (8.68) and (8.64) of the molar configuration entropy for an order-disorder transition in the Bragg-Williams approximation. Here $T_C = Z\omega/2k_B$. (Compare Fig. 8.10 in the text.)

(b) Equation (8.69) then gives the heat capacity in terms of S_C

$$C_V = T\frac{\partial S_C}{\partial T}$$

The molar heat capacity calculated from Fig. 8.4 using Eq. (8.69) is shown in Fig. 8.5.

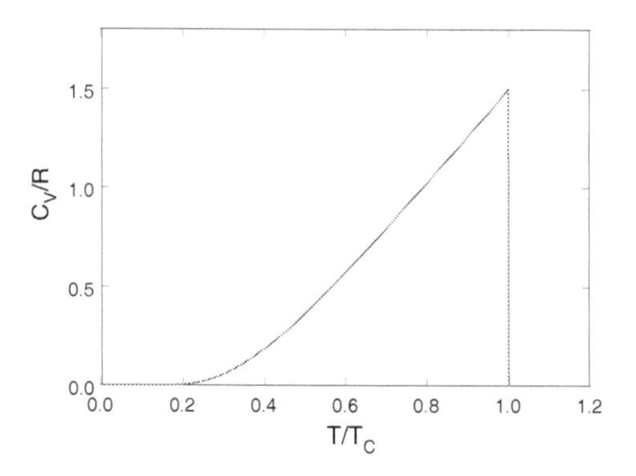

Fig. 8.5 Molar heat capacity obtained from Fig. 8.4 using Eq. (8.69). (Compare Fig. 8.11 in the text.)

Problem 8.10 Solve Eq. (8.110) numerically for the exact value of $k_{\rm B}T_C/J$ in the two-dimensional Ising model, and compare with the number quoted in Table 8.1 in the text.[8]

Solution to Problem 8.10

We want to solve Eq. (8.110)

$$1 - \sinh\left(\frac{2}{t}\right) = 0 \qquad ; \; t \equiv \frac{k_{\rm B}T_C}{J}$$

With the use of Mathcad 7, the numerical answer is

$$t = 2.2692$$

Therefore, the transition temperature is

$$k_{\rm B}T_C = 2.2692J$$

This is the result quoted in Table 8.1 in the text for the exact value of $k_{\rm B}T_C$ in the two-dimensional Ising model.

[8]Note $J \equiv \omega/2$.

Problem 8.11 (a) Solve Eq. (8.125) numerically, and verify the plot in Fig. 8.20 in the text of the magnetization per particle m as a function of T/T_C for the Ising model in MFT in any number of dimensions $d \geq 2$;

(b) What is the slope of this result at $T = T_C$?

(c) Locate the exact result for m in the two-dimensional Ising model, and compare with the result in (a).

Solution to Problem 8.11

(a) Below T_C, the magnetization per particle m as a function of T/T_C for the Ising model in MFT in any number of dimensions $d \geq 2$ is given by the solution to Eq. (8.125)

$$m = \tanh\left(\frac{T_C}{T}m\right)$$

The result, obtained using Mathcad 7, is shown in Fig. 8.6. This reproduces the plot in Fig. 8.20 in the text.

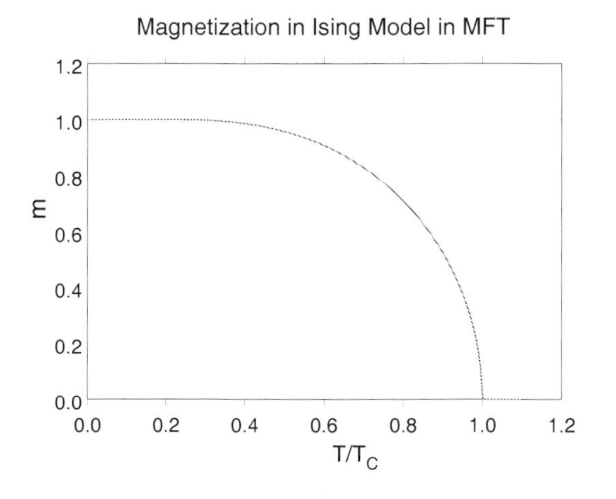

Fig. 8.6 Magnetization per system m for the Ising model in MFT in any number of dimensions $d \geq 2$ calculated from Eq. (8.125). Here, from Eq. (8.124), $k_B T_C = Jd$.

(b) It certainly appears that the magnetization comes in with infinite slope at $T = T_C$.

(c) Although analytic solutions exist for the 2-D Ising model [Onsager (1944); Yang (1952)], it is useful for present purposes to compare with some numerical results, for example [Kiel (2016)], where the magnetization m is calculated on a 64×64 lattice. The "exact" numerical result is squarer (less rounded) than that in Fig. 8.6, and there is, in addition, a high-T tail.[9]

Problem 8.12 (a) Solve the one-dimensional Ising model in the presence of an external magnetic field, which produces an additional interaction

$$H' = -\mu_0 B \sum_{\kappa=1}^{N} S_\kappa$$

(b) Show there is no phase transition to a ferromagnetic state.

Solution to Problem 8.12

(a) With the additional magnetic field, the canonical partition function in Eq. (8.85) is given by

$$(\text{P.F.}) = \sum_{S_1} \cdots \sum_{S_N} \exp\left[\frac{\omega}{2k_{\text{B}}T} \sum_{\kappa=1}^{N} S_\kappa S_{\kappa+1} + \frac{\mu_0 B}{k_{\text{B}}T} \sum_{\kappa=1}^{N} S_\kappa\right] \qquad ; \; S_\kappa = \pm 1$$

Extend the definition of the 2×2 matrix in Eq. (8.87) to

$$\langle S|P|S'\rangle \equiv \exp\left\{\frac{\omega S S'}{2k_{\text{B}}T} + \frac{\mu_0 B}{2k_{\text{B}}T}(S + S')\right\} \qquad ; \; (S, S') = \pm 1$$

which keeps it real and symmetric. Then

$$\underline{P} = \begin{bmatrix} e^{(\omega+2\mu_0 B)/2k_{\text{B}}T} & e^{-\omega/2k_{\text{B}}T} \\ e^{-\omega/2k_{\text{B}}T} & e^{(\omega-2\mu_0 B)/2k_{\text{B}}T} \end{bmatrix}$$

Equation (8.90) then becomes[10]

$$\begin{aligned}(\text{P.F.}) &= \sum_{S_1} \cdots \sum_{S_N} \prod_{\kappa=1}^{N} \exp\left[\frac{\omega}{2k_{\text{B}}T} S_\kappa S_{\kappa+1} + \frac{\mu_0 B}{2k_{\text{B}}T}(S_\kappa + S_{\kappa+1})\right] \\ &= \sum_{S_1} \cdots \sum_{S_N} \langle S_1|P|S_2\rangle\langle S_2|P|S_3\rangle \cdots \langle S_{N-1}|P|S_N\rangle\langle S_N|P|S_{N+1}\rangle\end{aligned}$$

The analysis then follows the text to obtain Eq. (8.96)

$$(\text{P.F.}) = \lambda_+^N \qquad ; \; N \to \infty$$

Thus we have to find the largest eigenvalue of $\underline{P}$.

[9]It is interesting, here, to compare to Fig. 8.25 in the text.
[10]Remember the p.b.c. [see Eq. (8.91)].

The eigenvalue Eq. (8.97) becomes

$$\det \begin{vmatrix} e^{(\omega+2\mu_0 B)/2k_BT} - \lambda & e^{-\omega/2k_BT} \\ e^{-\omega/2k_BT} & e^{(\omega-2\mu_0 B)/2k_BT} - \lambda \end{vmatrix} = 0$$

This gives

$$\lambda^2 - 2\lambda\, e^{\omega/2k_BT} \cosh\left(\frac{\mu_0 B}{k_BT}\right) + 2\sinh\left(\frac{\omega}{k_BT}\right) = 0$$

The largest eigenvalue then follows as[11]

$$\lambda_+(T,B) = e^{\omega/2k_BT} \cosh\left(\frac{\mu_0 B}{k_BT}\right) + \left[e^{\omega/k_BT} \cosh^2\left(\frac{\mu_0 B}{k_BT}\right) - 2\sinh\left(\frac{\omega}{k_BT}\right)\right]^{1/2}$$

The exact solution to the one-dimensional Ising model with the external field B is then, for large N,

$$\begin{aligned} A(T,B) &= -k_BT \ln(\text{P.F.}) \\ &= -Nk_BT \ln \lambda_+(T,B) \end{aligned}$$

(b) The magnetization is computed from the expression in part (a) as[12]

$$\mathcal{M} = k_BT \frac{\partial}{\partial B} \ln(\text{P.F.})$$

This gives

$$\mathcal{M} = N\mu_0 \sinh\left(\frac{\mu_0 B}{k_BT}\right) \times \frac{1}{\lambda_+(T,B)} \left\{ e^{\omega/2k_BT} + \frac{e^{\omega/k_BT}\cosh(\mu_0 B/k_BT)}{\left[e^{\omega/k_BT}\cosh^2(\mu_0 B/k_BT) - 2\sinh(\omega/k_BT)\right]^{1/2}} \right\}$$

As $B \to 0$, this reduces to

$$\begin{aligned} \mathcal{M} &\to N\mu_0 \sinh\left(\frac{\mu_0 B}{k_BT}\right) \left[\frac{1+e^{\omega/k_BT}}{1+e^{-\omega/k_BT}}\right] \qquad ; \; B \to 0 \\ &= N\mu_0 \sinh\left(\frac{\mu_0 B}{k_BT}\right) e^{\omega/k_BT} \end{aligned}$$

[11]Note that at $B = 0$, this reproduces Eq. (8.100) with $\lambda_+(T,0) = 2\cosh(\omega/2k_BT)$.

[12]Compare Prob. 7.20.

This vanishes for $B = 0$ at all temperatures. There is no residual magnetization at zero field at any T, and hence no phase transition to a ferromagnetic state.

Problem 8.13 Demonstrate that one can achieve a physical realization of the configuration in Fig. 8.18 in the text by placing the lattice on the surface of a *torus*.

Solution to Problem 8.13

Periodic boundary conditions in one dimension are achieved by putting the assembly around a cylinder. When you go around the cylinder, you come back where you started. Periodic boundary conditions in the second dimension can then be physically achieved by joining the ends of the cylinder and converting the cylinder into a *torus* (that is, a *doughnut*). Now when you go around in the second dimension, you again come back to where you started (see Fig. 8.7).[13]

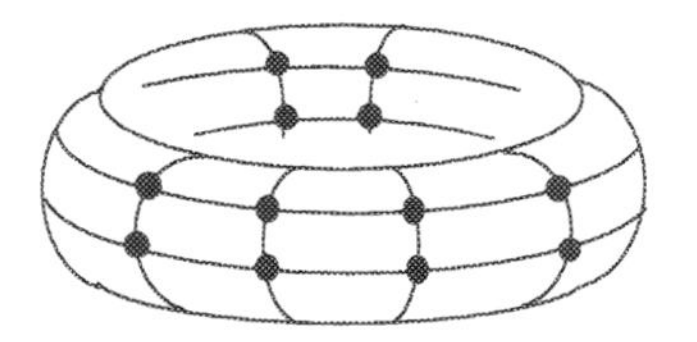

Fig. 8.7 Periodic boundary conditions in both directions achieved by putting the assembly on the surface of a torus.

Problem 8.14 (a) Show the following real orthogonal matrix satisfies $\underline{U}^T = \underline{U}^{-1}$

$$\underline{U} = \frac{1}{\sqrt{2}} \begin{pmatrix} 1 & 1 \\ 1 & -1 \end{pmatrix}$$

(b) Show $\underline{U}\,\underline{P}\,\underline{U}^{-1} = \underline{P}_D$, where $\underline{P}$ is the real symmetric matrix in Eq. (8.89), and $\underline{P}_D$ is the diagonal matrix in Eq. (8.93) with the real eigenvalues in Eq. (8.99).

[13] Compare Prob. 8.5.

Solution to Problem 8.14

(a) Consider $\underline{U}^T\underline{U}$

$$\underline{U}^T\underline{U} = \frac{1}{2}\begin{pmatrix} 1 & 1 \\ 1 & -1 \end{pmatrix}\begin{pmatrix} 1 & 1 \\ 1 & -1 \end{pmatrix} = \frac{1}{2}\begin{pmatrix} 2 & 0 \\ 0 & 2 \end{pmatrix} = \begin{pmatrix} 1 & 0 \\ 0 & 1 \end{pmatrix}$$

Therefore

$$\underline{U}^T = \underline{U}^{-1}$$

(b) Now work out $\underline{U}\,\underline{P}\,\underline{U}^{-1}$ where $\underline{P}$ is the real symmetric matrix in Eq. (8.89)

$$\begin{aligned}
\underline{U}\,\underline{P}\,\underline{U}^{-1} &= \frac{1}{2}\begin{pmatrix} 1 & 1 \\ 1 & -1 \end{pmatrix}\begin{pmatrix} e^{\omega/2k_{\rm B}T} & e^{-\omega/2k_{\rm B}T} \\ e^{-\omega/2k_{\rm B}T} & e^{\omega/2k_{\rm B}T} \end{pmatrix}\begin{pmatrix} 1 & 1 \\ 1 & -1 \end{pmatrix} \\
&= \frac{1}{2}\begin{pmatrix} 1 & 1 \\ 1 & -1 \end{pmatrix}\begin{pmatrix} e^{\omega/2k_{\rm B}T} + e^{-\omega/2k_{\rm B}T} & e^{\omega/2k_{\rm B}T} - e^{-\omega/2k_{\rm B}T} \\ e^{\omega/2k_{\rm B}T} + e^{-\omega/2k_{\rm B}T} & e^{-\omega/2k_{\rm B}T} - e^{\omega/2k_{\rm B}T} \end{pmatrix} \\
&= \begin{pmatrix} e^{\omega/2k_{\rm B}T} + e^{-\omega/2k_{\rm B}T} & 0 \\ 0 & e^{\omega/2k_{\rm B}T} - e^{-\omega/2k_{\rm B}T} \end{pmatrix}
\end{aligned}$$

Hence

$$\underline{U}\,\underline{P}\,\underline{U}^{-1} = \underline{P}_D$$

where $\underline{P}_D$ is the diagonal matrix in Eq. (8.93) with the real eigenvalues in Eq. (8.99).

Problem 8.15 (a) Consider the one-dimensional Ising model with periodic boundary conditions. Show

$$\sum_{\kappa=1}^{N} S_\kappa S_{\kappa+1} = N - 2N_{AB}$$

where N_{AB} is the number of unlike nearest-neighbor pairs;

(b) Show the mean energy of the assembly of spins is[14]

$$E = -\frac{\omega}{2}N + \omega\,\langle N_{AB}\rangle$$

where the mean number of unlike pairs at a temperature T is given by

$$\langle N_{AB}\rangle = \frac{\sum_{\{S\}} N_{AB}\, e^{-\omega N_{AB}/k_{\rm B}T}}{\sum_{\{S\}} e^{-\omega N_{AB}/k_{\rm B}T}}$$

Here the sum is over all spin configurations $\{S\} = (S_1, S_2, \cdots, S_N)$;

[14]As usual, $\langle E\rangle \equiv E$.

(c) Show the Bragg-Williams approximation for $\langle N_{AB}\rangle$ is

$$\overset{o}{N}_{AB} = \frac{N}{2}$$

(d) Use the result in Eq. (8.103) to show that for large N, the exact answer for $\langle N_{AB}\rangle$ is

$$\langle N_{AB}\rangle = \frac{N}{2}\left[1 - \tanh\left(\frac{1}{x}\right)\right] \qquad ; \; x \equiv \frac{2k_{\rm B}T}{\omega}$$

Solution to Problem 8.15

The hamiltonian for the one-dimensional Ising model is given in Eq. (8.81)

$$H_{\rm Ising} = -\frac{\omega}{2}\sum_{\kappa=1}^{N} S_\kappa S_{\kappa+1} \qquad ; \text{ Ising model}$$

where the spin on each site takes the values $S_k = \pm 1$.

(a) Consider the illustration of the chain in Fig. (8.14) in the text, and move down the chain. If we examine the relation to the next neighbor on the right, we will have included all nearest-neighbor pairs when we finish moving down the chain.[15] The total number of nearest-neighbor pairs in this one-dimensional problem is evidently just the number of spins N. If that neighbor is unlike, include a (-1) in the sum. If it is like, include a $(+1)$. Let the total number of unlike pairs in a given configuration be N_{AB}. The total number of like pairs is then $N - N_{AB}$, where N is fixed. The result for the required sum will therefore be

$$\sum_{\kappa=1}^{N} S_\kappa S_{\kappa+1} = (N - N_{AB}) - N_{AB} = N - 2N_{AB}$$

(b) The hamiltonian for the one-dimensional Ising model can then be re-written as

$$\begin{aligned} H_{\rm Ising} &= -\frac{\omega}{2}\sum_{\kappa=1}^{N} S_\kappa S_{\kappa+1} \qquad ; \text{ Ising model} \\ &= -\frac{\omega}{2}N + \omega\, N_{AB} \end{aligned}$$

[15]The p.b.c. ensure this works out right when we come to the end of the chain.

The mean energy is given by

$$\langle E\rangle = \frac{\sum_{\{S\}} H\, e^{-H/k_{\rm B}T}}{\sum_{\{S\}} e^{-H/k_{\rm B}T}}$$
$$= -\frac{\omega}{2}N + \omega\langle N_{AB}\rangle$$

where the mean number of unlike pairs at a temperature T is given by[16]

$$\langle N_{AB}\rangle = \frac{\sum_{\{S\}} N_{AB}\, e^{-\omega N_{AB}/k_{\rm B}T}}{\sum_{\{S\}} e^{-\omega N_{AB}/k_{\rm B}T}}$$

The sum is over all spin configurations $\{S\} = (S_1, S_2, \cdots, S_N)$

(c) As we move down the chain with a random distribution of spins, it is equally likely that the next spin will be like, or unlike. Hence the Bragg-Williams approximation for the mean number of unlike pairs $\langle N_{AB}\rangle$ is simply

$$\overset{o}{N}_{AB} = \frac{N}{2}$$

(d) The exact expression for the energy in the 1-D Ising model is given for large N in Eqs. (8.103)–(8.104)[17]

$$E = \frac{N\omega}{2}e(x)$$
$$e(x) = -\tanh\left(\frac{1}{x}\right) \qquad ; \; x = \frac{2k_{\rm B}T}{\omega}$$

It follows that for large N, the exact answer for $\langle N_{AB}\rangle$ is

$$\langle N_{AB}\rangle = \frac{N}{2}\left[1 - \tanh\left(\frac{1}{x}\right)\right] \qquad ; \; x \equiv \frac{2k_{\rm B}T}{\omega}$$

Note that at high T, this reduces to the Bragg-Williams value.

Problem 8.16 (a) Carry out a *numerical Monte Carlo* calculation of the quantity $\langle N_{AB}\rangle$ in Prob. 8.15(b) through the following series of steps:

- Generate a random spin configuration;
- Compute N_{AB} for this configuration;
- Add the appropriate contributions in the numerator and denominator;
- Repeat

[16]The term $e^{-\omega N/2k_{\rm B}T}$ factors and cancels in the ratio.

[17]Consistent with our usage, $\langle E\rangle \equiv E$ is now the thermodynamic energy.

(b) Start with $N = 20$. (What is the total number of spin configurations in this case?) Compute $\langle N_{AB} \rangle / N$ for $0 < x < 40$, where $x = 2k_{\rm B}T/\omega$. Be sure and include some points in the interval $0 < x < 2$. Keep as many configurations as you can in the Monte Carlo calculation;

(c) Make a plot comparing with Bragg-Williams and with the exact answer. Discuss the convergence of your numerical calculation with respect to the number of configurations employed, and with respect to N.[18]

Solution to Problem 8.16

(a) Monte Carlo calculations are made for modern computers. The basic idea is as follows: flip a coin and assign 1 for heads and 0 for tails. Add these up, and divide by the number of tosses. Plot the distribution of the values you obtain for many sequences of $\mathcal{N}$ flips. If $\mathcal{N}$ is very large, this distribution is strongly peaked at a mean value of 1/2, and the width of the distribution narrows as $1/\sqrt{\mathcal{N}}$. This means that for very large $\mathcal{N}$, the answer you get for any given sequence will almost certainly be 1/2.

We wish to calculate the mean number of unlike pairs $\langle N_{AB} \rangle$ at a temperature T in the one-dimensional Ising model. From Prob. 8.15, this is

$$\langle N_{AB} \rangle = \sum_{\{S\}} N_{AB}\, P_{N_{AB}}$$

$$P_{N_{AB}} \equiv \frac{e^{-\beta N_{AB}}}{\sum_{\{S\}} e^{-\beta N_{AB}}} \qquad ; \ \beta = \frac{\omega}{k_{\rm B}T}$$

We perform this calculation with a simple Mathcad 11 program. The program is structured as follows:

- Write a subroutine that generates a random distribution of spins on N_S sites, and then counts the number of unlike nearest neighbors N_{AB} with this distribution;[19]
- Now write a main program that, at each β,
  - Initializes a numerator v_0 and denominator v_1 at zero;
  - Uses the subroutine to generate a distribution and return a N_{AB};
  - Adds $N_{AB}e^{-\beta N_{AB}}$ to v_0 and $e^{-\beta N_{AB}}$ to v_1;
  - Repeats N_C times;
  - Returns v_0/v_1.

[18] Problems 8.16 and 8.18 are as long as you want to make them, but they are two problems with which the students (and the author!) have had the most fun.

[19] Remember the periodic boundary conditions. To avoid confusion, in this solution we label the number of sites by N_S and number of configurations by N_C.

For large enough N_C, this should converge to the answer.

(b) Consider $N_S = 20$. The exact number of configurations in this case is

$$2^{N_S} = 1.049 \times 10^6 \qquad ; \; N_S = 20$$

A Monte Carlo calculation of $\langle N_{AB} \rangle / N_S$ with $N_C = 10^3$ is shown as a function of $1/\beta = k_{\rm B}T/\omega$ in Fig. 8.8. This is compared with the exact answer for large N_S from Prob. 8.15

$$\frac{\langle N_{AB} \rangle}{N_S} = \frac{1}{2}\left[1 - \tanh\left(\frac{1}{x}\right)\right] \qquad ; \; x \equiv \frac{2k_{\rm B}T}{\omega}$$

The high-temperature behavior comes out remarkably well, but the Monte Carlo calculation begins to deviate from the exact result below $1/\beta \approx 1.5$.

The extension to more configurations with $N_C = 10^5$ is shown in Fig. 8.9. Here the agreement with the exact result indeed holds down to $1/\beta \approx 1.0$.

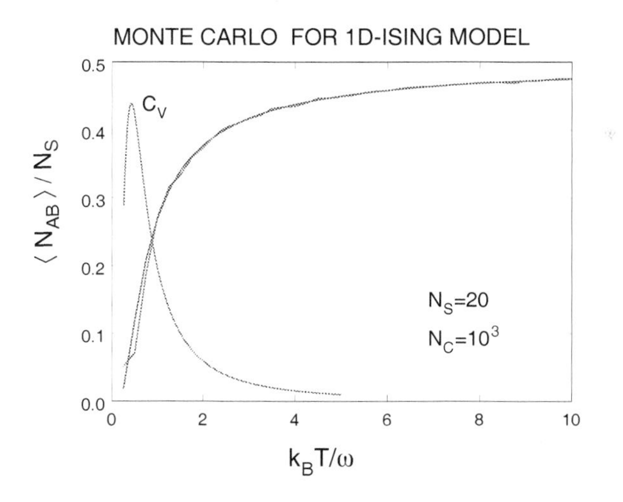

Fig. 8.8 Monte Carlo calculation of the mean number of unlike pairs $\langle N_{AB} \rangle / N_S$ as a function of temperature for the 1-D Ising model with $N_S = 20$ sites compared with the exact answer for large N_S. This calculation retains $N_C = 10^3$ configurations. The exact answer for the heat capacity $C_V/N_S k_{\rm B}$ is also shown (recall Fig. 8.16 in the text).

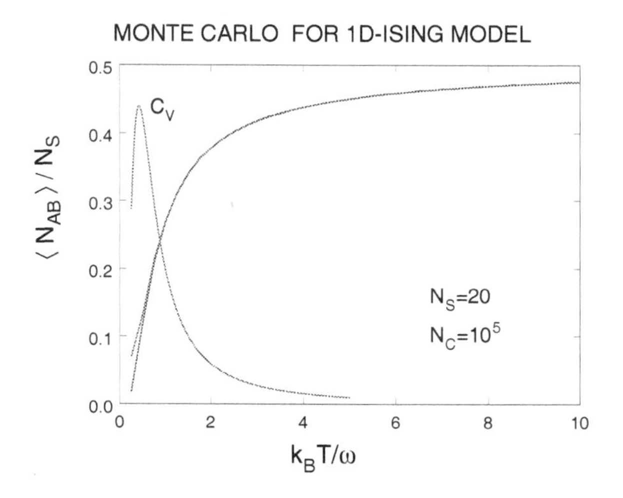

Fig. 8.9 Same as Fig. 8.8 retaining $N_C = 10^5$ configurations.

Note the Bragg-Williams result is given in Prob. 8.15 as the high-T limit

$$\frac{\langle N_{AB} \rangle}{N_S} = \frac{1}{2} \qquad ; \text{ Bragg-Williams}$$

Also shown in these figures is the exact result for the heat capacity (see Prob. 8.17 and the discussion there)

$$\frac{C_V}{N_S k_{\rm B}} = 2\,\frac{\partial}{\partial x}\frac{\langle N_{AB} \rangle}{N_S} = \left(\frac{1}{x^2}\right)\text{sech}^2\left(\frac{1}{x}\right) \qquad ; \; x \equiv \frac{2k_{\rm B}T}{\omega}$$

(c) The Monte Carlo calculation in Fig. 8.9 is extended to a larger number of sites with $N_S = 40$ in Fig. 8.10. In this case the total number of configurations is

$$2^{N_S} = 1.1 \times 10^{12} \qquad ; \; N_S = 40$$

Here the numerical calculation only samples a very small fraction of the possible configurations; however, the high-temperature behavior is again well reproduced. This Monte Carlo calculation begins to deviate badly below $1/\beta \approx 1.5$.

Once one has the program, it is fun to let it run to see what it takes to improve the convergence at low T for various N_S.

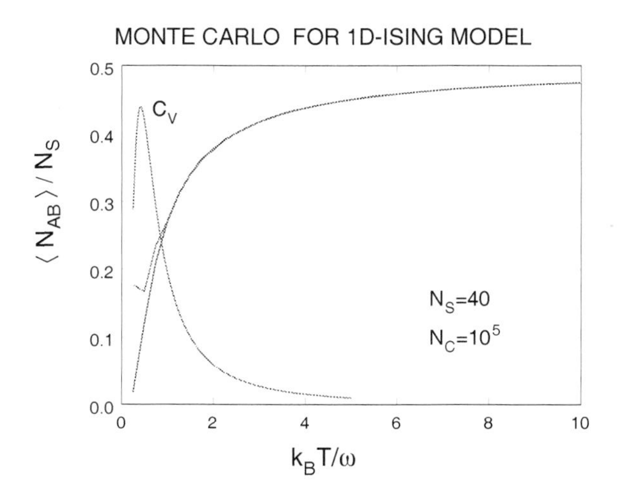

Fig. 8.10 Same as Fig. 8.9 with $N_S = 40$ sites.

Problem 8.17 (a) Show that the heat capacity in the one-dimensional Ising model is given for large N by

$$\frac{C_V}{Nk_{\rm B}} = 2\,\frac{\partial}{\partial x}\frac{\langle N_{AB}\rangle}{N} = \left(\frac{1}{x^2}\right)\text{sech}^2\left(\frac{1}{x}\right)$$

Include this quantity in your plot in Prob. 8.16(c);

(b) Explain why it is much more difficult to obtain the heat capacity with your Monte Carlo calculation.

Solution to Problem 8.17

(a) The heat capacity can be obtained by differentiating the energy in Prob. 8.15

$$E = -\frac{\omega}{2}N + \omega\langle N_{AB}\rangle$$
$$C_V = \left(\frac{\partial E}{\partial T}\right)_N = \omega\frac{\partial}{\partial T}\langle N_{AB}\rangle$$

Hence, with $x = 2k_{\rm B}T/\omega$,

$$\frac{C_V}{Nk_{\rm B}} = 2\,\frac{\partial}{\partial x}\frac{\langle N_{AB}\rangle}{N} \qquad ;\; x \equiv \frac{2k_{\rm B}T}{\omega}$$

From Prob. 8.15, the exact solution for large N for the mean number of unlike pairs is

$$\frac{\langle N_{AB}\rangle}{N} = \frac{1}{2}\left[1 - \tanh\left(\frac{1}{x}\right)\right]$$

Since $d\tanh u/du = \operatorname{sech}^2 u$, It follows that

$$\frac{C_V}{Nk_{\rm B}} = \left(\frac{1}{x^2}\right)\operatorname{sech}^2\left(\frac{1}{x}\right)$$

(b) The heat capacity involves a derivative

$$C_V = \text{Lim}_{\Delta T\to 0}\,\frac{1}{\Delta T}\left[E(T+\Delta T) - E(T)\right]$$

Any numerical uncertainty in calculating $E(T)$ is magnified in calculating the *difference* $[E(T+\Delta T) - E(T)]$ between neighboring points, particularly when $E(T)$ is small and changing rapidly.

Problem 8.18 An improvement of the Monte Carlo calculation is achieved with the use of the *Metropolis algorithm*, which provides increased convergence [Metropolis, Rosenbluth, Rosenbluth, Teller, and Teller (1953)]. Although the derivation goes beyond the scope of this work,[20] it is easy to implement. The summand in Prob. 8.15(b) involves a probability distribution

$$P(N_{AB}) = \frac{e^{-\omega N_{AB}/k_{\rm B}T}}{\sum_{\{S\}} e^{-\omega N_{AB}/k_{\rm B}T}}$$

The idea is to replace

$$\sum_{\{S\}} P(N_{AB}) \to \sum_{\{C\}}$$

where the sum now goes over all members of a *new ensemble of spin configurations*, with the number of members of the new ensemble in a region around N_{AB} reflecting the probability distribution $P(N_{AB})$. Then $\langle N_{AB}\rangle$, for example, is simply given by

$$\langle N_{AB}\rangle = \frac{\sum_{\{C\}} N_{AB}}{\sum_{\{C\}}}$$

The Metropolis algorithm provides a method for finding the new ensemble. Let the assembly *thermalize* for several rounds:[21]

[20]It is derived, for example, in [Walecka (2004)].

[21]This has been re-worded slightly for clarity (see the following solution).

(1) Start with a spin configuration $\{S\}$ with its N_{AB};
(2) Generate a new spin configuration by changing one or more spins;
(3) Compute the new $\tilde{N}_{AB}$ and the ratio

$$r = \frac{P(\tilde{N}_{AB})}{P(N_{AB})} = \frac{e^{-\omega \tilde{N}_{AB}/k_{\rm B}T}}{e^{-\omega N_{AB}/k_{\rm B}T}}$$

(4) If $r > 1$, retain the configuration, and go back to step (2);
(5) If $r < 1$, generate a random number R between $[0, 1]$. If $R \leq r$, retain the configuration; if not, discard it, retaining the original. Now go back to step (2).

The accumulated spin configurations then yield the desired ensemble. The numerical accuracy, of course, depends on the number of members of the ensemble.

Carry out the numerical calculation in Prob. 8.16 using the Metropolis algorithm.

Solution to Problem 8.18

The Metropolis algorithm is derived, for example, in [Walecka (2004)]. Here we present the results of a very simple Mathcad 11 program that incorporates this algorithm in computing $\langle N_{AB}\rangle$. There are three parts to the program:

(1) The first part generates a random distribution of spins on N_S sites, and counts the number N_{AB} of unlike nearest-neighbor pairs;[22]
(2) The second part applies the Metropolis algorithm. We start with a random configuration and corresponding N_{AB}. A new random distribution and corresponding $\tilde{N}_{AB}$ are then generated. The following ratio is computed

$$r = \frac{e^{-\omega \tilde{N}_{AB}/k_{\rm B}T}}{e^{-\omega N_{AB}/k_{\rm B}T}}$$

- If $r > 1$, the old configuration is replaced by the new one
- If $r < 1$, a random number R between $[0, 1]$ is generated
  - If $R \leq r$, the old configuration is replaced
  - If not, the old configuration is retained

[22] This is done here by generating a random number between $-1/2$ and $+1/2$ on each site, and then checking the sign of the product for neighboring sites as we move down the chain. At the end of the chain, periodic boundary conditions are applied.

This procedure is repeated N_T times to produce one member of a properly *thermalized* distribution;

(3) The quantity $\langle N_{AB}\rangle$ is then calculated as

$$\langle N_{AB}\rangle = \frac{1}{N_C}\sum_{\{C\}} N_{AB}$$

where the sum goes over N_C members of the thermalized distribution.

The resulting calculation of $\langle N_{AB}\rangle$ with $N_S = 40$ sites and $N_C = 10^4$ configurations, using $N_T = 10^3$ thermalization cycles, is shown as a function of $1/\beta = k_\mathrm{B}T/\omega$ in Fig. 8.11. The exact answer for large N_S is again indicated. This figure should be compared with the Monte Carlo results in Fig. 8.10. We make several comments:

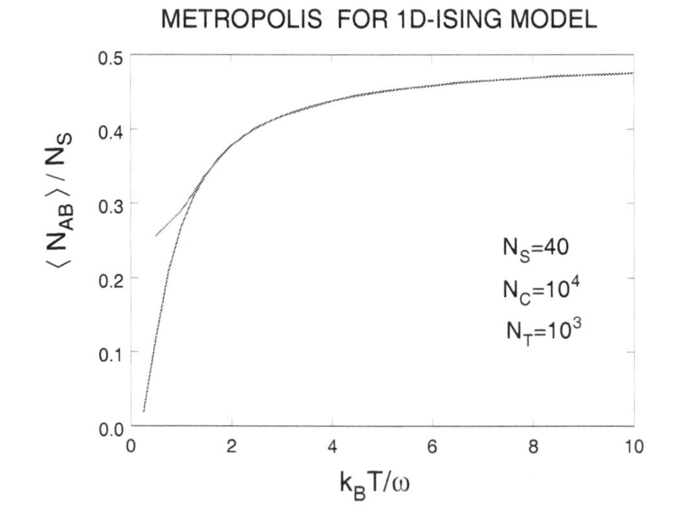

Fig. 8.11 Same calculation as in Fig. 8.10 using the Metropolis algorithm with $N_C = 10^4$ configurations. There are $N_T = 10^3$ thermalization cycles. Here $1/\beta = k_\mathrm{B}T/\omega$.

- These results employ only a tiny fraction of the total number of spin configurations $2^{N_S} = 1.1 \times 10^{12}$ with $N_S = 40$. Here only 10^4 configurations are used;
- A rapid approach to the exact answer is obtained at large $1/\beta$;
- This calculation reproduces the exact answer down to $1/\beta \approx 1$;

- More work is needed to get down to small $1/\beta$. To illustrate the magnitude of the problem, as $1/\beta \to 0$ we seek the ground state with all spins aligned (there are two of them), out of 1.1×10^{12} configurations;
- Once again, given the program, it is fun to investigate convergence with respect to (N_S, N_C, N_T)
- The present calculation for the 1-D Ising model using the Metropolis algorithm is neither elegant nor efficient. There is plenty of room for readers to do a better job on this one!

The following presents a simple example that demonstrates just how the Metropolis algorithm works.

****

(*Aside*) Although it takes us a little further afield, it is useful to have a simple solvable example where we can understand these numerical methods in more detail, and properly apply them. Particularly, since they are so widely used in today's physics.[23]

Consider the following mean value

$$\langle x \rangle = \frac{\int_0^1 x e^{-\beta x}\, dx}{\int_0^1 e^{-\beta x}\, dx} \equiv I(\beta)$$

where we define this ratio of integrals as $I(\beta)$. This expression can be evaluated analytically[24]

$$I(\beta) = \frac{1}{\beta} - \frac{1}{e^{\beta} - 1}$$

Write the integral as a finite sum with N terms and a uniform spacing dx along the x-axis. The spacing *dx factors and cancels in the ratio.* Thus we have

$$\langle x \rangle \sum_{i=1}^{N} e^{-\beta x_i} = \sum_{i=1}^{N} x_i e^{-\beta x_i}$$

[23]See, for example, the later discussion of lattice gauge theory

[24]Use

$$\int_0^1 e^{-\beta x}\, dx = \frac{1}{\beta}\left(1 - e^{-\beta}\right)$$

$$\int_0^1 x e^{-\beta x}\, dx = -\frac{d}{d\beta}\int_0^1 e^{-\beta x}\, dx = \frac{1}{\beta^2}\left(1 - e^{-\beta}\right) - \frac{e^{-\beta}}{\beta}$$

Let us now re-interpret this expression, as illustrated in Fig. 8.12.

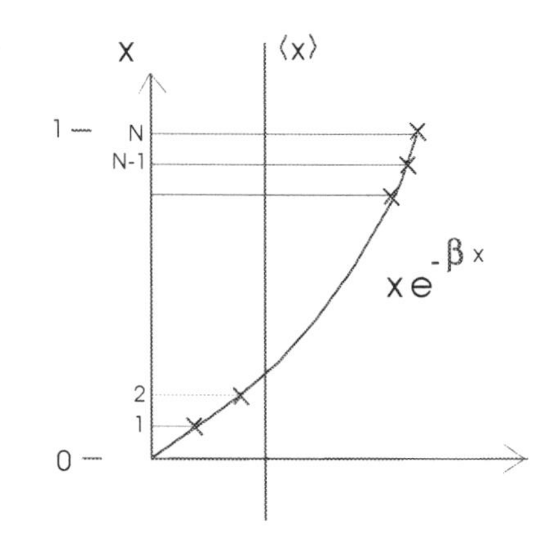

Fig. 8.12 Re-interpretation of the integral for the mean value $\langle x \rangle$. Basis of the Monte Carlo calculation.

The quantity $x_i e^{-\beta x_i}$ is the distance out to the curve $xe^{-\beta x}$ at the point i. We sum all the lengths, and then divide by the sum of the weighting lengths $e^{-\beta x_i}$ to get the mean value $\langle x \rangle$ for the curve $xe^{-\beta x}$. But now it is clear from this figure that in computing this mean value, there is no reason we need sample the curve at a set of points with uniform spacing. We could just as well sample it with a set of *randomly* distributed points

$$\langle x \rangle \sum_{\{N\}} e^{-\beta x_i} = \sum_{\{N\}} x_i e^{-\beta x_i}$$

where $\{N\}$ denotes *a set of N points randomly distributed along the x-axis in the interval* $[0, 1]$. In the limit $N \to \infty$, this should produce the same result for $\langle x \rangle$. This provides the basis for a Monte Carlo calculation of the mean value.

We use Mathcad 7 to write a simple Monte Carlo program $MC(\beta)$ that

- Initializes a numerator v_0 and denominator v_1 at zero;
- Generates a random number x on the interval $[0, 1]$;
- Adds $xe^{-\beta x}$ to v_0 and $e^{-\beta x}$ to v_1;
- Repeats N times;
- Returns v_0/v_1.

The result for $MC(\beta)$ with $N = 10^4$ is compared with the exact answer for $I(\beta)$ in Fig. 8.13. Within an error of $O(\sqrt{N}\,)$, they are clearly identical.

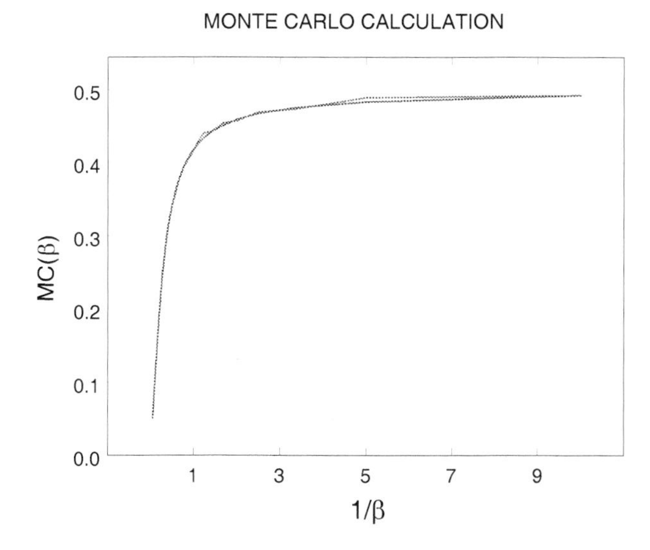

Fig. 8.13 Calculation of the mean value $I(\beta) \equiv \int_0^1 xe^{-\beta x}\,dx \;/\; \int_0^1 e^{-\beta x}\,dx$ using the Monte Carlo method $\text{MC}(\beta)$ with $N = 10^4$ points, compared with the exact answer.

The analysis can now be extended. Re-write the above as

$$\langle x \rangle = \sum_{\{N\}} x_i P(x_i, \beta)$$

$$P(x_i, \beta) \equiv \frac{e^{-\beta x_i}}{\sum_{\{N\}} e^{-\beta x_i}}$$

The quantity $P(x_i, \beta)$ has the interpretation as the fractional *probability* that with a distribution $e^{-\beta x_i}$, the point i at the position x_i contributes to the mean value. If the set $\{N\}$ of random points can be re-distributed to a *new* set $\{C\}$ that already reflects this probability distribution, then our expression for $\langle x \rangle$ can be written simply as

$$\langle x \rangle N = \sum_{\{C\}} x_i$$

The Metropolis algorithm provides a procedure for finding this new set of points.

We write a simple Mathcad 11 program that converts the initial set of random points $\{N\}$ into a new set $\{C\}$ distributed according to the probability distribution $P(x_i, \beta)$.

- Introduce a matrix $U_{i,j}$ where the row indicates the iteration and the column is the location of one of the N points;
- For $U_{1,j}$ start with a random distribution of points on the interval $[0,1]$;
- For each j and corresponding x_j, generate a new random position $\tilde{x}_j$;
- Compute

$$r = \frac{e^{-\beta \tilde{x}_j}}{e^{-\beta x_j}}$$

  - If $r > 1$, move the point;
  - If $r < 1$, generate a random number R between $[0,1]$;
    * If $R \leq r$, move the point;
    * If not, do not move the point;

- Repeat this process on the previous distribution NI times to *thermalize* it. The distribution $U_{NI,j}$ should then be stable under this procedure;[25]
- The new distribution of points $\{C\}$ is now given by $U_{NI,j}$

Figure 8.14 presents a histogram constructed from $U_{NI,j}/N$ with $NI = 20$ iterations, starting from a uniform, random distribution. The points are binned into twenty intervals on the x-axis between 0 and 1. Also shown in this figure is the correct continuum limit of this probability distribution

$$\frac{e^{-\beta x}\, dx}{\int_0^1 e^{-\beta x}\, dx} = \left[\frac{\beta}{(1-e^{-\beta})} e^{-\beta x}\right] \frac{1}{N_{\text{int}}}$$

Here we have used $dx = 1/N_{\text{int}}$, with $N_{\text{int}} = 20$. This histogram demonstrates that we have achieved the proper probability distribution.[26]

The Metropolis result for $\langle x \rangle$, which we refer to as MET(β), follows from this analysis as

$$\text{MET}(\beta) = \frac{1}{N} \sum_{j=1}^{N} U_{NI,j}$$

It is now clear from Fig. 8.14 why the Metropolis algorithm will produce the correct mean value. The number of points in each bin provides just the right amount of weighting to reproduce the integral. The advantage of the Metropolis algorithm is that it allows one to take into account important regions in x that might be missed in the random Monte Carlo sampling.

[25] See, for example, [Walecka (2004)].

[26] The Metropolis algorithm really works! (There is no significance to the fact that here $N_{\text{int}} = N_{NI}$.) Note the result is the same as that obtained from following each point N_{NI} times.

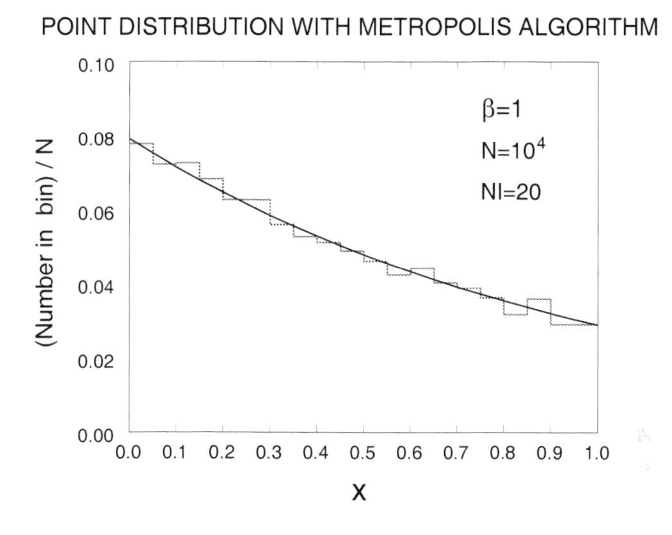

Fig. 8.14 Histogram of the point distribution $U_{NI,j}/N$ after application of the Metropolis algorithm with $\beta = 1$, $N = 10^4$, and $NI = 20$ iterations starting from a uniform, random distribution. There are twenty intervals in the histogram. Also shown is the correct continuum limit of this probability distribution (see text).

A comparison to the exact $I(\beta)$ of the Monte Carlo and Metropolis calculations with $N = 10^4$ is shown in Table 8.1. Here the two numerical methods give comparable results, very close to the correct answer for $\langle x \rangle$.

Table 8.1 Calculation of the mean value $I(\beta) \equiv \int_0^1 xe^{-\beta x}\,dx \;/\int_0^1 e^{-\beta x}\,dx$ using the Monte Carlo method MC(β) and the Metropolis algorithm MET(β) with $N = 10^4$ points.

β	$1/\beta$	I(β)	MC(β)	MET(β)
0.1	10	0.492	0.490	0.489
0.2	5	0.483	0.480	0.483
1	1	0.418	0.418	0.420
5	0.2	0.193	0.193	0.192
10	0.1	0.100	0.100	0.105

****

Problem 8.19 (a) Show that for imaginary time, the lagrangian density for the scalar field in Eq. (8.130) gives $\mathcal{L}(\mathbf{x}, \tau) \equiv [\mathcal{L}(x_\mu)]_{t=-i\tau} = -\mathcal{H}(\mathbf{x}, \tau)$, where $\mathcal{H}$ is the hamiltonian density;

(b) Since the total hamiltonian $H = \int d^3x\, \mathcal{H}$ is a constant of the motion, show that the exponential in Eqs. (8.131)–(8.133) is $e^{-\beta H}$.

Solution to Problem 8.19

(a) The lagrangian density for the scalar field with a non-linear self-coupling in Eq. (8.130) is[27]

$$\begin{aligned}\mathcal{L} &= -\frac{1}{2}\left[\left(\frac{\partial\phi}{\partial x_\mu}\right)^2 + m^2\phi^2\right] - \frac{\lambda}{4!}\phi^4 \\ &= \frac{1}{2}\left(\frac{\partial\phi}{\partial t}\right)^2 - \frac{1}{2}(\boldsymbol{\nabla}\phi)^2 - \frac{1}{2}m^2\phi^2 - \frac{\lambda}{4!}\phi^4\end{aligned}$$

In continuum mechanics, the canonical momentum density and hamiltonian density are given by [Fetter and Walecka (2003a)]

$$\begin{aligned}\pi &= \frac{\partial\mathcal{L}}{\partial(\partial\phi/\partial t)} = \frac{\partial\phi}{\partial t} \\ \mathcal{H} &= \pi\frac{\partial\phi}{\partial t} - \mathcal{L} \\ &= \frac{1}{2}\left(\frac{\partial\phi}{\partial t}\right)^2 + \frac{1}{2}(\boldsymbol{\nabla}\phi)^2 + \frac{1}{2}m^2\phi^2 + \frac{\lambda}{4!}\phi^4\end{aligned}$$

It is now evident that if we evaluate the lagrangian density for imaginary time, we obtain the negative of the hamiltonian density

$$\mathcal{L}(\mathbf{x},\tau) \equiv [\mathcal{L}(x_\mu)]_{t=-i\tau} = -\mathcal{H}(\mathbf{x},\tau)$$

(b) The weighting in the partition function in Eqs. (8.131)–(8.133) is then[28]

$$\begin{aligned}Z &= \int \mathcal{D}(\phi)\, e^{-S(\beta,0)} \\ S(\beta,0) &= -\int_0^\beta d\tau \int d^3x\, \mathcal{L}(\mathbf{x},\tau) = \int_0^\beta d\tau \int d^3x\, \mathcal{H}(\mathbf{x},\tau)\end{aligned}$$

[27] Recall that here $x_\mu = (\mathbf{x}, ict)$, and repeated Greek indicies are summed from 1 to 4. We remind the reader that for this final set of LGT problems, we use units where $\hbar = c = 1$, and now $\beta \equiv +1/k_{\rm B}T$.

[28] In more detail, $[\mathcal{L}(\phi, \boldsymbol{\nabla}\phi, \partial\phi/\partial t)]_{t=-i\tau} = \mathcal{L}(\phi, \boldsymbol{\nabla}\phi, i\partial\phi/\partial\tau) = -\mathcal{H}(\phi, \boldsymbol{\nabla}\phi, \partial\phi/\partial\tau)$. The path integral for Z is then analyzed by dividing the integration region for the action S into steps in τ, and integrating over the field at each τ (see [Walecka (2004)]).

The total hamiltonian, here the total energy, is a constant of the motion

$$\int d^3x\, \mathcal{H}(\mathbf{x},\tau) = H = \text{constant}$$

Therefore

$$S(\beta,0) = H\int_0^\beta d\tau = \beta H$$

Thus, for this theory of a scalar field with a non-linear self-coupling, the partition function is

$$Z = \int \mathcal{D}(\phi)\, e^{-\beta H} \qquad\qquad ; \ \beta = \frac{1}{k_{\rm B}T}$$

Problem 8.20 Use the two-dimensional version of Stokes' theorem

$$\oint_C \mathbf{v}\cdot d\mathbf{l} = \int_A (\boldsymbol{\nabla}\times\mathbf{v})\cdot d\mathbf{S}$$

to verify the continuum limit in Eq. (8.144) in $1+1$ dimensions ($d=2$).[29] Show that in this case σ must be chosen so that $2\sigma = 1/e_0^2a^2$.

Solution to Problem 8.20

The configuration for U(1) lattice gauge theory in $1+1$ dimensions is illustrated in Figs. 8.21–8.22 in the text. The field variables are associated with the links as shown in Eq. (8.137)[30]

$$U_{\rm link} \equiv e^{ie_0A_\mu(x)[x(j)-x(i)]_\mu}$$

The contribution to the action from these link variables is taken as the product around a plaquette [see Eqs. (8.138)]

$$\begin{aligned}
U_\Box &\equiv U_1U_2U_3U_4 \\
S_\Box &\equiv 2\sigma\,(1-\mathrm{Re}\,U_\Box)
\end{aligned}$$

[29]The continuum limit here is obtained by keeping the leading term in the expansion of the first of Eqs. (8.137), and assuming the rest of the series is well behaved in the limit $a\to 0$.

[30]Recall Eqs. (8.135); note that x_μ and $A_\mu(x)$ are now real.

Here 2σ is a constant, which can depend on the charge e_0 and lattice spacing a. The total action is obtained from the sum over all plaquettes [see Eq. (8.141)]

$$S = \sum_{\Box} S_{\Box}$$

Let the lattice spacing $a \to 0$. Then $U_{\Box}$ can be written as the infinitesimal form of the two-dimensional line integral

$$U_{\Box} = \exp\left\{ie_0 \oint_C \mathbf{A} \cdot d\mathbf{l}\right\}$$

where the integral goes around the edge of the plaquette. Now substitute this expression in $S_{\Box}$, and expand in powers of a. The leading term is

$$S_{\Box} = \frac{2\sigma e_0^2}{2!}\left(\oint_C \mathbf{A} \cdot dl\right)^2 \qquad ; \ a \to 0$$

Use Stokes' theorem, as stated in the problem. This becomes[31]

$$\begin{aligned} S_{\Box} &= \frac{2\sigma e_0^2}{2!}\left[\int_S (\mathbf{\nabla} \times \mathbf{A}) \cdot d\mathbf{S}\right]^2 \qquad ; \ a \to 0 \\ &= \frac{2\sigma e_0^2}{2!}\left[a^2 (\mathbf{\nabla} \times \mathbf{A})_3\right]^2 \\ &= \frac{2\sigma e_0^2 a^2}{2!}\left[a^2\left(\frac{\partial A_2}{\partial x_1} - \frac{\partial A_1}{\partial x_2}\right)^2\right] \end{aligned}$$

This result is, in turn,

$$S_{\Box} = \frac{2\sigma e_0^2 a^2}{4}\left[a^2 F_{\mu\nu} F_{\mu\nu}\right] \qquad ; \ a \to 0$$

where repeated Greek indices are again summed from 1 to 2.

Now choose

$$2\sigma = \frac{1}{e_0^2 a^2}$$

In the limit $a \to 0$, the total action then becomes

$$S = \int d\tau dx \, \frac{1}{4} F_{\mu\nu} F_{\mu\nu} \qquad ; \ a \to 0$$

This is the euclidean form of the continuum action in Eq. (8.144) for $d = 2$.

[31] Here the third axis is normal to the page, and pointing out.

Problem 8.21 Extend Prob. 8.20 to $3+1$ dimensions ($d = 4$). Show that in this case $2\sigma = 1/e_0^2$. Note that here σ is independent of a. (This problem requires more effort.)

Solution to Problem 8.21

With the solution to Prob. 8.20 in front of us, this problem is not so difficult. Divide the sum over plaquettes into a sum over the unit cell, as illustrated in Fig. 8.24 in the text, and then sum over all the unit cells

$$S_{\Box}^{(i)} = \sum_{\text{unit cell}} S_{\Box}$$

$$S = \sum_i S_{\Box}^{(i)}$$

The analysis in Prob. 8.20 then applies to each plaquette in the unit cell. Hence, in $3+1$ dimensions, with $d = 4$, one has

$$S_{\Box}^{(i)} = \frac{2\sigma e_0^2}{4}\left[a^4 F_{\mu\nu}^{(i)} F_{\mu\nu}^{(i)}\right] \qquad ; \ a \to 0$$

where repeated Greek indices are now summed from 1 to 4.

In this case, since a^4 is the volume element of the unit cell, we now choose

$$2\sigma = \frac{1}{e_0^2}$$

Then, as $a \to 0$,

$$S = \sum_i S_{\Box}^{(i)} = \int d\tau\, d^3x\, \frac{1}{4} F_{\mu\nu} F_{\mu\nu} \qquad ; \ a \to 0$$

This is the euclidean form of the continuum action in Eq. (8.144) for $d = 4$.

Problem 8.22 (a) Expand the exponential in Eq. (8.143), and derive the following result for U(1) lattice gauge theory in the strong-coupling limit

$$\langle S_{\Box} \rangle = 2\sigma \qquad ; \ \sigma \to 0$$

(b) Calculate the next term, and show

$$\langle S_{\Box} \rangle = 2\sigma - \frac{1}{2}(2\sigma)^2 \qquad ; \ \sigma \to 0$$

Solution to Problem 8.22

(a) The partition function for U(1) lattice gauge theory is given by Eq. (8.143)

$$Z \equiv \prod_l \int_0^{2\pi} \frac{d\phi_l}{2\pi} \exp\left[-\sum_{\Box} S_{\Box}\right]$$

where the plaquette action is that in Eq. (8.155)

$$\begin{aligned} S_{\Box} &= \sigma\left\{[1-(U_{\Box})_{\hookrightarrow}] + [1-(U_{\Box})_{\hookleftarrow}]\right\} \\ &= \sigma\left\{[1-(U_{\Box})_{\hookrightarrow}] + \left[1-(U_{\Box})^{\star}_{\hookrightarrow}\right]\right\} \end{aligned}$$

Here $U_{\Box}$ is given in terms of link variables by Eq. (8.139) [see Figs. 8.21–8.22 in the text]

$$U_{\Box} = e^{i\phi_1} e^{i\phi_2} e^{-i\phi_3} e^{-i\phi_4}$$

The expectation value of the plaquette action is

$$\langle S_{\Box}\rangle = \frac{1}{Z}\prod_l \int_0^{2\pi} \frac{d\phi_l}{2\pi} S_{\Box} \exp\left[-\sum_{\Box} S_{\Box}\right]$$

Substitution of the expression for $S_{\Box}$ gives

$$\begin{aligned} \langle S_{\Box}\rangle &= 2\sigma + \mathcal{R} \\ \mathcal{R} &= -\sigma\frac{1}{Z}\prod_l \int_0^{2\pi} \frac{d\phi_l}{2\pi} [(U_{\Box})_{\hookrightarrow} + (U_{\Box})^{\star}_{\hookrightarrow}] \exp\left[-\sum_{\Box} S_{\Box}\right] \end{aligned}$$

It is shown below that $\mathcal{R}$ is of $O(\sigma^2)$ as $\sigma \to 0$, and this then gives the first result that $\langle S_{\Box}\rangle \to 2\sigma$ as $\sigma \to 0$.

(b) The integral over the link variables satisfies Eq. (8.154)

$$\frac{1}{2\pi}\int_0^{2\pi} d\phi\, e^{in\phi} = \delta_{n,0}$$

Hence the integral over unpaired links vanishes, and the only way to get a non-vanishing result is to have paired links with $e^{i\phi}e^{-i\phi}$.

Now expand the exponential in $\mathcal{R}$ for small σ. The leading contribution comes from the first-order term in the expansion, and to obtained fully-paired links, it is only the term in $S_\Box$ that contributes[32]

$$\begin{aligned}\mathcal{R} &\approx -\sigma^2 \prod_l \int_0^{2\pi} \frac{d\phi_l}{2\pi} [(U_\Box)_{\hookrightarrow} + (U_\Box)^\star_{\hookrightarrow}][(U_\Box)_{\hookrightarrow} + (U_\Box)^\star_{\hookrightarrow}] \\ &= -2\sigma^2\end{aligned}$$

where we have paired $(U_\Box)_{\hookrightarrow}\,(U_\Box)^\star_{\hookrightarrow}$ to get a non-zero result. Hence

$$\langle S_\Box \rangle = 2\sigma - \frac{1}{2}(2\sigma)^2 \qquad ; \; \sigma \to 0$$

This is Eq. (8.156).

Problem 8.23 (a) Solve Eq. (8.150) numerically, and verify Eq. (8.151); (b) Reproduce the MFT result in Fig. 8.25 in the text.

Solution to Problem 8.23

(a) The relation in Eq. (8.150) satisfied by the magnetization m in U(1) mean-field theory is

$$m = \frac{\int_0^{2\pi} d\phi \, \cos\phi \, e^{2z\sigma m^3 \cos\phi}}{\int_0^{2\pi} d\phi \, e^{2z\sigma m^3 \cos\phi}} \qquad ; \text{ mean-field theory}$$

A solution was found to this relation using Mathcad 7, provided $\sigma > \sigma_C$ where

$$z\sigma_C = 2.7878 \qquad ; \text{ critical value}$$

(b) Figure 8.15 shows the magnetization m in mean-field theory for U(1) lattice gauge theory calculated from Eq. (8.150) using Mathcad 7. What is plotted is m^4 vs. σ_C/σ, where from Eqs. (8.151)–(8.152), $z\sigma_C = 2.7878$ with $z = d - 1$. This is the MFT result in Fig. 8.25 in the text. Note that in this form, we have a universal curve, valid for any d.

Problem 8.24 Use the results in Fig. 8.26 in the text to add the following row to Table 8.2 in the text: $[6, 0.5576, 0.33\cdots]$.

Solution to Problem 8.24

The mean-field theory result for the "transition temperature" in six

[32] To this order, the partition function in the denominator can be replaced by $Z = 1$.

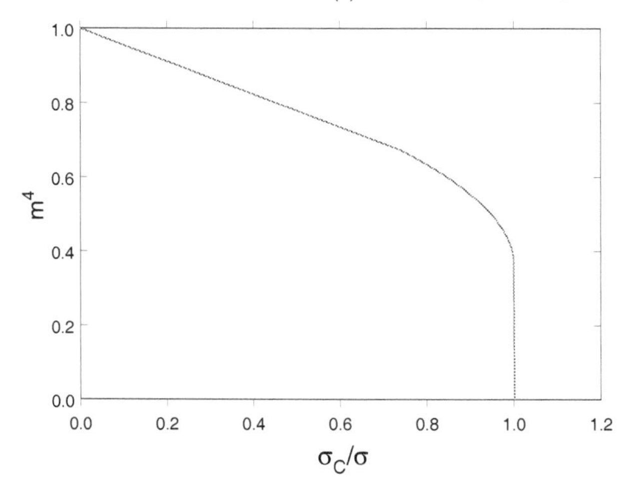

Fig. 8.15 Magnetization m in mean-field theory from Eq. (8.150) for U(1) lattice gauge theory. What is plotted is m^4 vs. σ_C/σ, where from Eqs. (8.151)–(8.152), $z\sigma_C = 2.7878$ with $z = d - 1$. (See Fig. 8.25 in the text.)

dimensions follows from Eq. (8.153)

$$\sigma_C = \frac{1}{5}(2.7878) = 0.5576 \qquad ; \text{ MFT}$$

From Fig. 8.26 in the text, the value obtained by [Barmore (1999)] from a numerical calculation on a 5^6 lattice is

$$\sigma_C \approx \frac{1}{2}(0.65\cdots) = 0.33\cdots \qquad ; \text{ numerical}$$

Thus we can extend Table 8.2 in the text to Table 8.2 below. Note the gradual convergence of the MFT values to the exact result.

Table 8.2 "Transition temperature" $\sigma_C = \beta_{\text{eff}}^C/2$ for U(1) lattice gauge theory in mean-field theory. The exact result for $d = 4$ was calculated on a 5^4 lattice by [Dubach (2004)], and the result for $d = 6$ on a 5^6 lattice by [Barmore (1999)].

Dimension d	σ_C(MFT)	Exact Result
2	2.7878	no phase transition
4	0.9293	0.4975
6	0.5576	$0.33\cdots$

Problem 8.25 Verify the thermodynamic relations in Eqs. (8.160)–(8.161) for $\beta = 1/k_{\rm B}T$.

Solution to Problem 8.25

The relations in Eqs. (8.160) are

$$A = -\frac{1}{\beta}\ln{(\text{P.F.})} \qquad ; \; \beta = \frac{1}{k_{\rm B}T}$$
$$E = \frac{\partial}{\partial\beta}(\beta A)$$

For $\beta = 1/k_{\rm B}T$, use

$$\frac{\partial}{\partial\beta} = \left(\frac{d\beta}{dT}\right)^{-1}\frac{\partial}{\partial T} = -k_{\rm B}T^2\frac{\partial}{\partial T}$$

Hence the above relations become

$$A = -k_{\rm B}T\ln{(\text{P.F.})}$$
$$E = -T^2\frac{\partial}{\partial T}\left(\frac{A}{T}\right)$$

These reproduce Eqs. (5.1) and (5.7) for the Helmholtz free energy and thermodynamic energy, respectively.

Problem 8.26 (a) Show it is no loss of generality to assume $b_0 = 1$ in Eq. (8.157);

(b) Construct the $[1, 1]$ Padé approximant for $c_0 + c_1\beta + c_2\beta^2$.

Solution to Problem 8.26

(a) The Padé approximant to a power series is defined in Eq. (8.157)

$$\sum_{i=0}^{n} c_i\beta^i = \frac{\sum_{j=0}^{p} a_j\beta^j}{\sum_{k=0}^{q} b_k\beta^k} \qquad ; \text{ Padé approximant } [p, q]$$
$$p + q = n$$

The r.h.s. is the ratio of two series. Divide the numerator and denominator by b_0. The new Padé approximant than has $b_0 = 1$.

(b) Express the given series as a $[1, 1]$ Padé approximant

$$c_0 + c_1\beta + c_2\beta^2 \approx \frac{a_0 + a_1\beta}{1 + b_1\beta}$$

An expansion of the r.h.s. through $O(\beta^2)$ then allows us to match the coefficients[33]

$$c_0 + c_1\beta + c_2\beta^2 = a_0[1 - b_1\beta + (b_1\beta)^2] + a_1\beta(1 - b_1\beta) + \cdots$$

Hence

$$\begin{aligned} c_0 &= a_0 \\ c_1 &= a_1 - a_0 b_1 \\ c_2 &= a_0 b_1^2 - a_1 b_1 \end{aligned}$$

The solution to these equations is

$$a_0 = c_0 \qquad ; \; b_1 = -\frac{c_2}{c_1} \qquad ; \; a_1 = c_1 - \frac{c_0 c_2}{c_1}$$

Note this Padé approximant has a singularity at $\beta = c_1/c_2$.

Problem 8.27 (a) Use the results in Prob. 8.22 to show that in the strong-coupling limit

$$E_\Box = 1 - \frac{1}{2}(2\sigma) + \cdots \qquad ; \text{ strong-coupling}$$

Compare with the results in Fig. 8.26 in the text for small $\beta_{\rm eff} = 2\sigma$;

(b) Solve Eq. (8.150) numerically for m in six dimensions ($d = 6$, $z = 5$), and calculate the MFT value of

$$E_\Box = (1 - m^4) \qquad ; \text{ MFT}$$

Compare with the results in Fig. 8.26 in the text for large $\beta_{\rm eff} = 2\sigma$, and extend the figure in $\beta_{\rm eff}$.

Solution to Problem 8.27

(a) From Eq. (8.158) and Prob. 8.22, we have in the strong-coupling limit[34]

$$E_\Box = \frac{1}{2\sigma}\langle S_\Box \rangle = 1 - \frac{1}{2}(2\sigma) + \cdots \qquad ; \text{ strong-coupling}$$

(b) The magnetization m in mean-field theory calculated from Eq. (8.150) for U(1) lattice gauge theory in six dimensions is shown in

[33]Use $1/(1-x) = 1 + x + x^2 + \cdots$.
[34]For $E_\Box$, see Prob. 8.25.

Fig. 8.16. What is plotted is m^4 vs. σ_C/σ, where from Eq. (8.151), $5\sigma_C = 2.7878$.

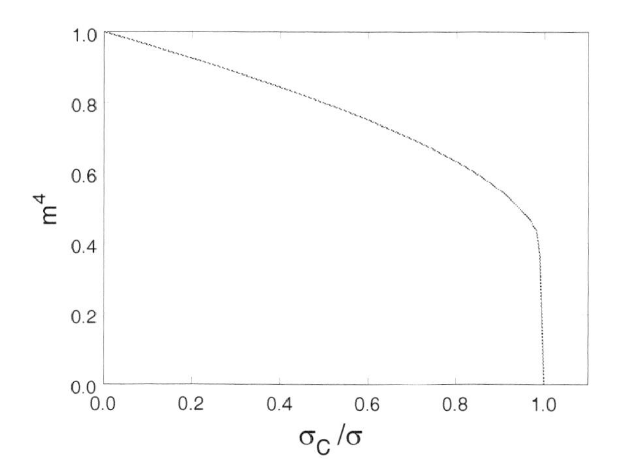

Fig. 8.16 Magnetization m in mean-field theory from Eq. (8.150) for U(1) lattice gauge theory in six dimensions. What is plotted is m^4 vs. σ_C/σ, where from Eq. (8.151), $5\sigma_C = 2.7878$. (Compare Fig. 8.25 in the text.)

The leading strong-coupling and mean-field results for the plaquette energy for U(1) LGT with $d = 6$ are shown in Fig. 8.17

$$
\begin{aligned}
E_\Box &\to 1 - \frac{1}{2}(2\sigma) && ; \text{ strong-coupling} && ; \sigma \to 0 \\
E_\Box &\to 1 - m^4 && ; \text{ mean-field} && ; \sigma \to \infty
\end{aligned}
$$

Here m is the solution to Eq. (8.150) with $z = 5$.

These asymptotic approximations should be compared with the full numerical calculations on a 5^6 lattices in Fig. 8.26 in the text, which uses an improved axial-gauge MFT and exhibits a first-order phase transition (note $2\sigma = \beta$).

Problem 8.28 Since the link variable in Eq. (8.137) is periodic in the phase ϕ_l, show the measure in Eq. (8.142) is gauge-invariant.

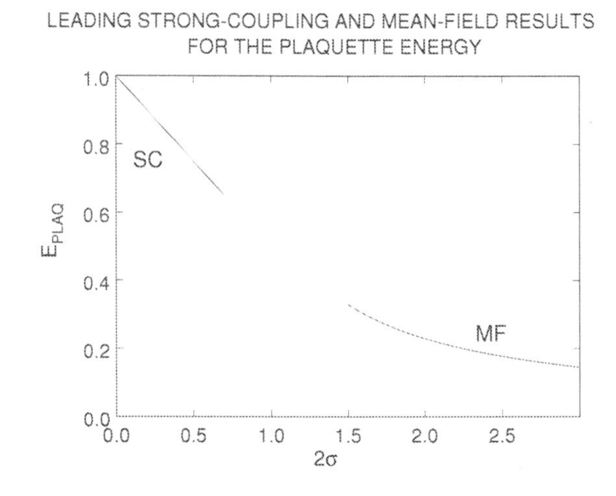

Fig. 8.17 Leading strong-coupling $E_\square = 1-(2\sigma)/2$ and mean-field $E_\square = 1-m^4$ results for the plaquette energy for U(1) LGT with $d=6$. Here m is the solution to Eq. (8.150) with $z=5$. This should be compared with Fig. 8.26 in the text (note $2\sigma = \beta$).

Solution to Problem 8.28

The link variable in Eq. (8.137) is given by

$$\begin{aligned} U_{\text{link}} &= e^{ie_0 A_\mu(x)[x(j)-x(i)]_\mu} \\ &\equiv e^{i\phi_l} \end{aligned}$$

Make a gauge transformation

$$A_\mu \to A_\mu + \frac{\partial \Lambda}{\partial x_\mu}$$

Then as the lattice spacing $a \to 0$,

$$[x(j)-x(i)]_\mu \frac{\partial \Lambda}{\partial x_\mu} \to \Lambda(x_j) - \Lambda(x_i)$$

Therefore, under a gauge transformation,

$$\begin{aligned} U_{\text{link}} &\to U_{\text{link}}\, e^{ie_0[\Lambda(x_j)-\Lambda(x_i)]} \\ &\equiv e^{i(\phi_l+\delta_l)} \qquad ; \ \delta_l = e_0[\Lambda(x_j)-\Lambda(x_i)] \end{aligned}$$

The measure is defined in Eq. (8.142)

$$\int [dA_\mu] \to \int_0^{2\pi} \frac{d\phi_l}{2\pi}$$

Then with a gauge transformation

$$\int_0^{2\pi} \frac{d\phi_l}{2\pi} F[e^{i\phi_l}] \to \int_0^{2\pi} \frac{d\phi_l}{2\pi} F[e^{i(\phi_l+\delta_l)}]$$
$$= \int_{\delta_l}^{2\pi+\delta_l} \frac{d\phi_l'}{2\pi} F[e^{i\phi_l'}] \qquad ; \ \phi_l' = \phi_l + \delta_l$$

Now write

$$\int_{\delta_l}^{2\pi+\delta_l} \frac{d\phi_l'}{2\pi} F[e^{i\phi_l'}] = \int_0^{2\pi} \frac{d\phi_l'}{2\pi} F[e^{i\phi_l'}]$$
$$+ \int_{2\pi}^{2\pi+\delta_l} \frac{d\phi_l'}{2\pi} F[e^{i\phi_l'}] - \int_0^{\delta_l} \frac{d\phi_l'}{2\pi} F[e^{i\phi_l'}]$$

The two terms in the second line cancel since the integrand is periodic in ϕ_l with period 2π. Hence, under a gauge transformation, the chosen measure leads to a gauge-invariant result

$$\int_0^{2\pi} \frac{d\phi_l}{2\pi} F[e^{i\phi_l}] \to \int_0^{2\pi} \frac{d\phi_l'}{2\pi} F[e^{i\phi_l'}]$$

Problem 8.29 At high temperature and pressure, nuclear matter dissolves into a *quark-gluon plasma*. Model this as a collection of free, massless, quarks, antiquarks, and gluons (bosons). Assume degeneracies of (γ_Q, γ_G) respectively. Use $H - \mu B$ in defining the (G.P.F.), where B is the baryon number, and recall that quarks and antiquarks carry baryon number $(1/3, -1/3)$.

(a) Show the gluon chemical potential must vanish;[35]

(b) Show the parametric equation of state is given by (here $\omega_k \equiv |\mathbf{k}|c$)

$$\frac{E}{V} = \frac{\gamma_Q}{(2\pi)^3} \int d^3k \, \hbar\omega_k \left[\frac{1}{e^{(\hbar\omega_k - \mu/3)/k_B T} + 1} + \frac{1}{e^{(\hbar\omega_k + \mu/3)/k_B T} + 1} \right]$$
$$+ \frac{\gamma_G}{(2\pi)^3} \int d^3k \frac{\hbar\omega_k}{e^{\hbar\omega_k/k_B T} - 1}$$
$$\frac{B}{V} = \frac{1}{3} \frac{\gamma_Q}{(2\pi)^3} \int d^3k \left[\frac{1}{e^{(\hbar\omega_k - \mu/3)/k_B T} + 1} - \frac{1}{e^{(\hbar\omega_k + \mu/3)/k_B T} + 1} \right]$$

(c) What is the pressure?

[35] Compare Prob. 7.28(a). In the hot, dense plasma, radiation by the quarks (and gluons) provides a mechanism for changing gluon number.

Solution to Problem 8.29

(a) The number of gluons is not conserved. It then follows exactly as in Eqs. (7.29)–(7.35) that the gluon chemical potential $\mu_{\rm gl}$ must vanish

$$\mu_{\rm gl} = 0 \qquad ; \; \lambda_{\rm gl} = e^{\mu_{\rm gl}/k_{\rm B}T} = 1$$

(b) We then define the grand partition function as

$$({\rm G.P.F.}) = \sum_{\{n\}} e^{-H/k_{\rm B}T}\, e^{B\mu/k_{\rm B}T}$$

where the hamiltonian and baryon number are given by [compare Eqs. (7.7)]

$$H = \sum_i (n_i\varepsilon_i + \bar{n}_i\varepsilon_i)_{\rm quark} + \sum_i (n_i\varepsilon_i)_{\rm gluon}$$
$$B = \frac{1}{3}\sum_i (n_i - \bar{n}_i)_{\rm quark}$$

Here we observe that the quarks (with occupation numbers n_i) carry baryon number $1/3$, while the antiquarks (with occupation numbers $\bar{n}_i$) carry $-1/3$. It follows that for a gas of massless gluons (bosons), quarks, and antiquarks (fermions) the grand partition function is determined as in Eqs. (7.11)–(7.15) to be

$$({\rm G.P.F.}) = \prod_i \left(\frac{1}{1 - e^{-\hbar\omega_k/k_{\rm B}T}}\right) \times \prod_i \left[1 + e^{-(\hbar\omega_k - \mu/3)/k_{\rm B}T}\right] \prod_i \left[1 + e^{-(\hbar\omega_k + \mu/3)/k_{\rm B}T}\right]$$

The energy is then given by Eq. (7.19)[36]

$$E = k_{\rm B}T^2 \frac{\partial}{\partial T} \ln({\rm G.P.F.})$$

The sum over states becomes

$$\sum_i \rightarrow \frac{\gamma_G V}{(2\pi)^3} \int d^3k \qquad ; \text{ gluons}$$
$$\sum_i \rightarrow \frac{\gamma_Q V}{(2\pi)^3} \int d^3k \qquad ; \text{ quarks}$$

[36]Remember (V, λ) are kept fixed here. Note $\lambda = e^{\mu/3k_{\rm B}T}$ is the absolute activity of the quarks.

where (γ_G, γ_Q) are the respective degeneracies.[37] The energy follows as

$$\frac{E}{V} = \frac{\gamma_Q}{(2\pi)^3} \int d^3k\, \hbar\omega_k \left[\frac{1}{e^{(\hbar\omega_k - \mu/3)/k_\mathrm{B}T} + 1} + \frac{1}{e^{(\hbar\omega_k + \mu/3)/k_\mathrm{B}T} + 1} \right] + \frac{\gamma_G}{(2\pi)^3} \int d^3k \frac{\hbar\omega_k}{e^{\hbar\omega_k/k_\mathrm{B}T} - 1}$$

It is evident from the above that the thermodynamic energy $E(V, T, \lambda)$ is the thermal average of the hamiltonian. The corresponding thermodynamic baryon number $B(V, T, \lambda)$ is the thermal average of the baryon number given by [compare Eq. (7.16)]

$$\begin{aligned} B &= k_\mathrm{B}T \frac{\partial}{\partial\mu} \ln(\mathrm{G.P.F.}) \\ &= \frac{1}{3} \frac{\partial}{\partial \ln\lambda} \ln(\mathrm{G.P.F.}) \qquad ; \ \lambda = e^{\mu/3k_\mathrm{B}T} \\ &= \frac{1}{3} \lambda \frac{\partial}{\partial\lambda} \ln(\mathrm{G.P.F.}) \end{aligned}$$

This gives

$$\frac{B}{V} = \frac{1}{3} \frac{\gamma_Q}{(2\pi)^3} \int d^3k \left[\frac{1}{e^{(\hbar\omega_k - \mu/3)/k_\mathrm{B}T} + 1} - \frac{1}{e^{(\hbar\omega_k + \mu/3)/k_\mathrm{B}T} + 1} \right]$$

Note the relative minus sign for the quarks and antiquarks in the baryon number; in contrast, both quarks and antiquarks contribute with positive sign in the energy.

(c) The pressure follows directly from the (G.P.F.)

$$PV = k_\mathrm{B}T \ln(\mathrm{G.P.F.})$$

Partial integrations for the bosons and fermions as in Eqs. (7.45)– (7.47) and Prob. 7.13 then give for this relativistic gas

$$PV = \frac{1}{3} E$$

We note that quantum chromodynamics (QCD), the theory of the strong interactions of quarks and gluons, is *asymptotically free* (see the following problem). This provides some justification for the current model.

[37] For the massless, spin-one gluons, $\gamma_G = 2\,(\text{helicities}) \times 8\,(\text{colors})$; while in the nuclear domain of massless, spin-one-half, (u, d) quarks, $\gamma_Q = 2\,(\text{helicities}) \times 3\,(\text{colors}) \times 2\,(\text{flavors})$.

Problem 8.30 In QED, the sum of leading logarithms gives a dependence of the fine structure constant on momentum transfer of[38]

$$\alpha(q^2) = \frac{\alpha}{1-(\alpha/3\pi)\ln(q^2/M_e^2)} \qquad ; \; q^2 \gg M_e^2$$

where M_e is the inverse Compton wavelength of the electron. Suppose the lattice spacing in U(1) lattice gauge theory is related to the momentum transfer by $a^2 = 1/q^2$. Show that with the above expression, a singularity is reached in the coupling constant *before* one gets to the continuum limit.

In asymptotically-free theories, the sign in the denominator of the corresponding expression is reversed.

Solution to Problem 8.30

It is evident that the denominator in the expression for $\alpha(q^2)$ vanishes, and hence $\alpha(q^2)$ has a singularity, when

$$\ln\left(\frac{q^2}{M_e^2}\right) = \frac{3\pi}{\alpha}$$
$$\frac{q^2}{M_e^2} = e^{3\pi/\alpha}$$

With a fine-structure constant of $\alpha = e^2/\hbar c = 1/137.0$, and electron inverse Compton wavelength of $1/M_e = \hbar/m_e c = 3.862 \times 10^{-11}\,\text{cm}$, this gives

$$\frac{q^2}{M_e^2} = e^{1291} = 10^{561}$$
$$q \approx 10^{291}\,\text{cm}^{-1}$$

This corresponds to a distance

$$\frac{1}{q} \approx 10^{-291}\,\text{cm}$$

Several comments:

- This is a ridiculously small distance. QED, the most accurate physical theory we have, has only been tested down to distances $\approx 10^{-16}\,\text{cm}$;
- However, if $\alpha(q^2)$ is the running coupling constant in U(1) lattice gauge theory, and if the relevant momentum transfer is related to the lattice spacing by $q^2 = 1/a^2$, then it is not possible to go to the continuum

[38] See, for example, [Walecka (2010)].

limit $a^2 \to 0$ since one encounters a singularity in the running coupling constant before reaching this limit;[39]

- Thus, whatever U(1) lattice gauge theory is, and whatever interesting properties it has, it is not a valid discretization of QED;
- It was a transforming breakthrough to realize that in a non-abelian theory such as quantum chromodynamics (QCD), the sign in the denominator of the running coupling constant $g(q^2)$ is *reversed*, and one can smoothly go to the continuum limit. In fact, in this limit, the running coupling constant goes to zero! This is known as asymptotic freedom [Gross and Wilczek (1973); Politzer (1973)];
- Quantum chromodynamics (QCD) *is* the modern theory of the strong interactions;
- In the case of QCD, lattice gauge theory provides a valid discretization of the theory [Wilson (1974)], and large-scale numerical calculations can represent the physical world;
- It is safe to say that lattice-gauge-theory calculations today form a dominant part of nuclear theory.[40]

[39] Note that in U(1) lattice gauge theory, the coupling constant gets *stronger* as $a^2 \to 0$.
[40] See [Walecka (2004)].

Appendix A

Non-Equilibrium Statistical Mechanics

Problem A.1 (a) Show explicitly that if the one-body distribution function has the form $f(H) = e^{-H/k_{\rm B}T}$, with $H = E$, the Boltzmann collision term in Eq. (A.18) vanishes;

(b) Then show explicitly that if $f(H) = e^{-H/k_{\rm B}T}$, with a one-body hamiltonian of the form in Eq. (A.19), the one-body distribution function in Eq. (A.20) satisfies

$$\frac{\partial f}{\partial t} = \boldsymbol{\nabla}_q U \cdot \boldsymbol{\nabla}_p f - \mathbf{v} \cdot \boldsymbol{\nabla}_q f = 0$$

Hence conclude that at a given point in phase space, f is independent of time.

Solution to Problem A.1

(a) If the one-body distribution function is of the form $f = f(H) = f(E)$, then the collision term in the Boltzmann Eq. (A.18) is proportional to

$$f_1'f_2' - ff_2 = f(E_1')f(E_2') - f(E)f(E_2)$$

With the given explicit form $f(E) = e^{-E/k_{\rm B}T}$, this becomes

$$f(E_1')f(E_2') - f(E)f(E_2) = e^{-(E_1'+E_2')/k_{\rm B}T} - e^{-(E+E_2)/k_{\rm B}T}$$

Energy is conserved in the two-body collision [see Eq. (A.25)]

$$E + E_2 = E_1' + E_2'$$

It follows that

$$f_1'f_2' - ff_2 = 0$$

Hence the Boltzmann collision term $(\partial f/\partial t)_{\text{collision}}$ in Eq. (A.18) *vanishes.*

(b) Consider the one-body problem in a mean field with the hamiltonian of Eq. (A.19)

$$H = \frac{\mathbf{p}^2}{2m} + U(\mathbf{q})$$

Suppose the one-body distribution function has the form

$$f(H) = e^{-H/k_B T}$$

implied by the vanishing of the Boltzmann collision term $(\partial f/\partial t)_{\text{collision}} = 0$ [see part (a)]. Then we have explicitly

$$\begin{aligned}\boldsymbol{\nabla}_q U \cdot \boldsymbol{\nabla}_p f - \mathbf{v} \cdot \boldsymbol{\nabla}_q f &= \left[\boldsymbol{\nabla}_q U(\mathbf{q}) \cdot \frac{\mathbf{p}}{m} - \frac{\mathbf{p}}{m} \cdot \boldsymbol{\nabla}_q U(\mathbf{q})\right] \frac{df(H)}{dH} \\ &= 0\end{aligned}$$

where we have used $\mathbf{v} = \mathbf{p}/m$. Hence Eq. (A.20) implies

$$\frac{\partial f}{\partial t} = \boldsymbol{\nabla}_q U \cdot \boldsymbol{\nabla}_p f - \mathbf{v} \cdot \boldsymbol{\nabla}_q f = 0$$

This confirms the equilibrium condition in Eq. (A.9) that the one-body distribution function at a given point in phase space is independent of time, even in the presence of a mean field.

Problem A.2 It was shown in the text that the one-body Fermi distribution in Eq. (A.37) leads to a vanishing of the collision term in the Nordheim-Uehling-Uhlenbeck equation. Show that with the one-body hamiltonian of Eq. (A.19), it is then still true that

$$\frac{\partial f}{\partial t} = \boldsymbol{\nabla}_q U \cdot \boldsymbol{\nabla}_p f - \mathbf{v} \cdot \boldsymbol{\nabla}_q f = 0$$

Solution to Problem A.2

The one-body Fermi distribution in Eq. (A.37) is

$$f(H) = \frac{1}{e^{\beta(\mu - H)} + 1} \equiv \frac{1}{D(H)}$$

where we now use $E \equiv H$. The one-body hamiltonian of Eq. (A.19) is

$$H = \frac{\mathbf{p}^2}{2m} + U(\mathbf{q})$$

Then, exactly as in Prob. (A.1),

$$\nabla_q U \cdot \nabla_p f - \mathbf{v} \cdot \nabla_q f = \left[\nabla_q U(\mathbf{q}) \cdot \frac{\mathbf{p}}{m} - \frac{\mathbf{p}}{m} \cdot \nabla_q U(\mathbf{q})\right] \frac{df(H)}{dH}$$
$$= 0$$

Hence Eq. (A.20), with a vanishing Nordheim-Uehling-Uhlenbeck collision term $(\partial f/\partial t)_{\text{collision}} = 0$, again implies

$$\frac{\partial f}{\partial t} = \nabla_q U \cdot \nabla_p f - \mathbf{v} \cdot \nabla_q f = 0$$

which is the desired equilibrium condition in Eq. (A.42).

Problem A.3 Consider two non-relativistic, equal-mass, non-identical particles interacting through a zero-range potential $V = \lambda\delta^{(3)}(\mathbf{x})$, where the relative and center-of mass (C-M) coordinates are given by $\mathbf{x} = \mathbf{x}_1 - \mathbf{x}_2$ and $\mathbf{R} = (\mathbf{x}_1 + \mathbf{x}_2)/2$.[1]

(a) Use Fermi's Golden Rule, and show the cross section for scattering into a region d^3p_1' about $\mathbf{p}_1'$ in the scattering process $\mathbf{p}_1 + \mathbf{p}_2 \to \mathbf{p}_1' + \mathbf{p}_2'$ is[2]

$$d\sigma_{fi} = \frac{\lambda^2}{v_{12}} \frac{2\pi}{\hbar} \delta(E_1 + E_2 - E_1' - E_2') \frac{d^3p_1'}{(2\pi\hbar)^3}$$

(b) Verify that overall momentum is conserved, and that the quantization volume drops out of this expression.

(c) Write the result in (a) as $\sigma d^3p_1'/(2\pi\hbar)^3$, and hence identify σv_{12}

$$d\sigma_{fi} \equiv \sigma \frac{d^3p_1'}{(2\pi\hbar)^3}$$

$$\sigma v_{12} = \frac{2\pi\lambda^2}{\hbar} \delta(E_1 + E_2 - E_1' - E_2')$$

(d) Use the result in (a) to derive the following expression for the differential cross section in the C-M system

$$\left(\frac{d\sigma}{d\Omega}\right)_{\text{CM}} = |f|^2 \qquad ; \; f = \frac{2\mu\lambda}{4\pi\hbar^2}$$

where μ is the reduced mass. Verify that this expression has the correct dimensions.

---

[1]This problem is longer, but central to the arguments in appendix A.
[2]See, for example, [Walecka (2008)].

Note that here σv_{12} depends only on $E_1 + E_2 = E'_1 + E'_2$, and the differential cross section in the C-M system is independent of scattering angle.

Solution to Problem A.3

(a) From footnote 4 in section A.1.2, the quantum mechanical expressions ("Golden Rule") for the rate and cross section for the scattering process in the second of Figs. A.2 in the text are given by

$$R_{fi} = \frac{2\pi}{\hbar}\delta(E_i - E_f)|\langle f|H'|i\rangle|^2$$
$$\sigma_{fi} = \frac{R_{fi}}{\text{Flux}}$$

The initial and final wave functions in the C-M system are given by

$$\psi_i = \frac{1}{V}e^{i\mathbf{P}\cdot\mathbf{X}/\hbar}\,e^{i\mathbf{p}\cdot\mathbf{x}/\hbar} \qquad ; \ \psi_f = \frac{1}{V}e^{i\mathbf{P}'\cdot\mathbf{X}/\hbar}\,e^{i\mathbf{p}'\cdot\mathbf{x}/\hbar}$$

where $(\mathbf{X}, \mathbf{x})$ are the C-M and relative coordinates as defined in the problem, and $(\mathbf{P}, \mathbf{p})$ are the total and relative momenta. The matrix element of the interaction is then given by

$$\langle f|H'|i\rangle = \frac{1}{V^2}\int d^3X \int d^3x\, e^{-i\mathbf{P}'\cdot\mathbf{X}/\hbar}\,e^{-i\mathbf{p}'\cdot\mathbf{x}/\hbar}\,\lambda\delta^{(3)}(\mathbf{x})\,e^{i\mathbf{P}\cdot\mathbf{X}/\hbar}\,e^{i\mathbf{p}\cdot\mathbf{x}/\hbar}$$

With p.b.c., this is

$$\langle f|H'|i\rangle = \frac{\lambda}{V}\delta_{\mathbf{p}_1+\mathbf{p}_2,\,\mathbf{p}'_1+\mathbf{p}'_2}$$

In summing over the final states in a small phase-space volume about $(\mathbf{p}'_1,\ \mathbf{p}'_2)$, the Kronecker delta can be employed to give

$$\sum_{\mathbf{p}'_1}\sum_{\mathbf{p}'_2}\delta^2_{\mathbf{p}_1+\mathbf{p}_2,\,\mathbf{p}'_1+\mathbf{p}'_2} = \sum_{\mathbf{p}'_1}\sum_{\mathbf{p}'_2}\delta_{\mathbf{p}_1+\mathbf{p}_2,\,\mathbf{p}'_1+\mathbf{p}'_2} = \sum_{\mathbf{p}'_1}$$
$$\to \frac{Vd^3p'_1}{(2\pi\hbar)^3} \qquad ; V \to \infty$$

The relative flux in the C-M system is

$$\text{Flux} = \frac{1}{V}v_{12}$$

Hence, the cross section is

$$d\sigma_{fi} = \frac{\lambda^2}{v_{12}}\frac{2\pi}{\hbar}\delta(E_1 + E_2 - E'_1 - E'_2)\frac{d^3p'_1}{(2\pi\hbar)^3}$$

which is the stated answer.

(b) The Dirac delta-function in the energy in the rate, and the Kronecker delta-function in the momentum in the matrix element, ensure that energy and momentum are conserved in the scattering process. Furthermore, it is clear from the derivation in part (a) that the fictitious quantization volume V cancels from the result for $d\sigma_{fi}$.

(c) If the result in (a) is written as $\sigma d^3p_1'/(2\pi\hbar)^3$, the quantity σv_{12} can be identified as[3]

$$\begin{aligned} d\sigma_{fi} &\equiv \sigma \frac{d^3p_1'}{(2\pi\hbar)^3} \\ \sigma v_{12} &= \frac{2\pi\lambda^2}{\hbar}\delta(E_1 + E_2 - E_1' - E_2') \end{aligned}$$

(d) With energy conservation in the C-M system, and μ the reduced mass, one can write

$$\begin{aligned} p_1' &= p' = p \\ d^3p_1' &= p'^2 dp' d\Omega' \\ E' &= E = \frac{p'^2}{2\mu} \\ v_{12} &= \frac{p}{\mu} = \frac{p'}{\mu} \end{aligned}$$

where (p, p') are the magnitudes of the initial and final relative momenta, and (E, E') are the total initial and final energies. In addition

$$\frac{dE'}{dp'} = \frac{p'}{\mu}$$

The cross section in part (a), integrated over dE', can then be written

$$\frac{d\sigma}{d\Omega'} = \frac{\lambda^2}{(p'/\mu)} \frac{2\pi}{\hbar} \left(\frac{dp'}{dE'}\right) \frac{p'^2}{(2\pi\hbar)^3}$$

[3]Without reference to the C-M system, it is still true that

$$\begin{aligned} \langle f|H'|i\rangle &= \frac{1}{V^2}\int d^3x_1 \int d^3x_2\, e^{-i\mathbf{p}_1'\cdot\mathbf{x}_1/\hbar}\, e^{-i\mathbf{p}_2'\cdot\mathbf{x}_2/\hbar}\, \lambda\delta^{(3)}(\mathbf{x}_1 - \mathbf{x}_2)\, e^{i\mathbf{p}_1\cdot\mathbf{x}_1/\hbar}\, e^{i\mathbf{p}_2\cdot\mathbf{x}_2/\hbar} \\ &= \frac{\lambda}{V}\delta_{\mathbf{p}_1+\mathbf{p}_2,\,\mathbf{p}_1'+\mathbf{p}_2'} \end{aligned}$$

and the relative flux is v_{12}/V. Hence, the derived relation for the rate σv_{12} holds in *any* frame.

The differential cross section in the C-M system therefore becomes

$$\left(\frac{d\sigma}{d\Omega}\right)_{\text{CM}} = |f|^2 \qquad ; \; f = \frac{2\mu\lambda}{4\pi\hbar^2}$$

Since λ has dimensions $[EL^3]$, and the dimensions of $2\mu/\hbar^2$ are $[E^{-1}L^{-2}]$, the dimension of the scattering amplitude f is properly $[L]$.

We note that here σv_{12} depends only on $E_1 + E_2 = E_1' + E_2'$, and the differential cross section in the C-M system is independent of scattering angle.

Problem A.4 Consider the thermodynamics of a uniform medium at temperature $T = 0$. Start from the Gibbs free energy.

(a) Show

$$\mathcal{H} = E + PV = N\mu$$
$$\mu = \left(\frac{\partial \mathcal{H}}{\partial N}\right)_P$$

where $\mathcal{H}$ is the *enthalpy* (recall Prob. 1.8), and μ is the chemical potential;

(b) Suppose the assembly is self-bound and in equilibrium with $P = 0$. Show the energy per system can be written in terms of the energy change upon insertion of a system as

$$e \equiv \frac{E}{N} = \left(\frac{\partial E}{\partial N}\right)_{P=0}$$

This is the Hugenholtz-Van Hove theorem.

Solution to Problem A.4

(a) The Gibbs free energy is defined in Eq. (1.21)

$$G \equiv E + PV - TS \qquad ; \text{ Gibbs free energy}$$

It satisfies Eq. (1.26)

$$dG = -SdT + VdP + \mu dN$$

where μ is again the chemical potential. Therefore

$$\mu = \left(\frac{\partial G}{\partial N}\right)_{T,P}$$

It is shown in Prob. 5.14 that the Gibbs free energy is the chemical potential per system

$$G(T, P, N) = N\mu$$

Suppose the assembly is in its ground state at zero temperature. Then the Gibbs free energy reduces to the enthalpy $\mathcal{H}$ of Prob. 1.8

$$G = E + PV = \mathcal{H} \qquad ; T = 0$$

At $T = 0$, the previous relations thus become

$$\mathcal{H} = E + PV = N\mu \qquad ; T = 0$$
$$\mu = \left(\frac{\partial \mathcal{H}}{\partial N}\right)_P$$

(b) Suppose, furthermore, that the assembly is self-bound and in equilibrium with $P = 0$. Then the enthalpy reduces to the energy

$$\mathcal{H} = E \qquad ; P = 0$$

The relations in part (a) then take the form

$$\mu = \frac{E}{N} = \left(\frac{\partial E}{\partial N}\right) \qquad ; (T, P) = 0$$

The energy per particle of a self-bound system at vanishing (T, P) is the change in energy as the particle number is changed. This is the Hugenholtz-Van Hove theorem.

Problem A.5 (a) Assume the uniform assembly in Prob. A.4 has an energy per system of the form $E/N = e(n)$ with $n = N/V$. Suppose it is under a non-zero pressure P. Verify the equation of state is given by

$$P(n) = n^2 \frac{de(n)}{dn} \qquad ; n = \frac{N}{V}$$

(b) Suppose one has a Fermi gas with a repulsive two-body interaction, held together under pressure. In the Hartree approximation, the single-particle potential $U(n)$, and energy per system in the assembly $e_{\text{Hartree}}(n)$, are given by[4]

$$U(n) = \tilde{V}(0)n$$

$$e_{\rm Hartree}(n) = \frac{3}{5}\varepsilon_F(n) + \frac{1}{2}\tilde{V}(0)n$$

Here $\tilde{V}(0)$ is the volume integral of the two-body potential, and $\varepsilon_F(n)$ is the Fermi energy in the second of Eqs. (7.122). Show that the corresponding pressure is given by

$$P(n) = P_F(n) + \frac{1}{2}nU(n)$$

where $P_F(n)$ is the Fermi gas pressure in the third of Eqs. (7.122). In this case, a measurement of the single-particle potential $U(n)$ yields the equation of state $P(n)$.

Solution to Problem A.5

(a) Write

$$E = Ne(n) \qquad ; \; n = \frac{N}{V}$$

Now derive the pressure from Eq. (1.17) at $T = 0$

$$\begin{aligned} P &= -\left(\frac{\partial E}{\partial V}\right)_N \\ &= N\frac{de(n)}{dn}\left(\frac{N}{V^2}\right) \\ &= n^2\frac{de(n)}{dn} \end{aligned}$$

(b) Suppose, as given, that one has a Fermi gas with a repulsive two-body interaction, held together under pressure. In the Hartree approximation, the single-particle potential $U(n)$, and energy per system in the assembly $e_{\rm Hartree}(n)$, are given by

$$U(n) = \tilde{V}(0)n$$

$$e_{\rm Hartree}(n) = \frac{3}{5}\varepsilon_F(n) + \frac{1}{2}\tilde{V}(0)n$$

[4]See, for example, [Fetter and Walecka (2003)]. This is another illustration of density functional theory.

Here $\tilde{V}(0)$ is the volume integral of the two-body potential, and $\varepsilon_F(n)$ is the Fermi energy in the second of Eqs. (7.122)

$$\varepsilon_{\rm F}(n) = \frac{\hbar^2}{2m}\left(\frac{6\pi^2}{g_s}\right)^{2/3} n^{2/3}$$

The pressure is then computed from the expression in part (a) as

$$\begin{aligned} P &= n^2 \frac{de_{\rm Hartree}(n)}{dn} \\ &= n^2 \left[\frac{2}{5}\frac{\varepsilon_F(n)}{n} + \frac{1}{2}\tilde{V}(0)\right] \\ &= \frac{2}{5} n\,\varepsilon_F(n) + \frac{1}{2} n^2 \tilde{V}(0) \end{aligned}$$

This gives

$$P(n) = P_F(n) + \frac{1}{2} n U(n)$$

where $P_F(n)$ is the Fermi gas pressure in the third of Eqs. (7.122), and the single-particle potential $U(n)$ is given above. In this case, a measurement of the single-particle potential $U(n)$ yields the equation of state $P(n)$.

Bibliography

Amit, D. J., (2005). *Field Theory: The Renormalization Group and Critical Phenomena, 3rd ed.*, World Scientific Publishing Company, Singapore

Amore, P., and Walecka, J. D., (2013). *Introduction to Modern Physics: Solutions to Problems*, World Scientific Publishing Company, Singapore

Amore, P., and Walecka, J. D., (2014). *Topics in Modern Physics: Solutions to Problems*, World Scientific Publishing Company, Singapore

Amore, P., and Walecka, J. D., (2015). *Advanced Modern Physics: Solutions to Problems*, World Scientific Publishing Company, Singapore

Barmore, B., (1999). *Acta. Phys. Polon.* **B30**, 1055

Bohr, A., and Mottelson, B. R., (1975). *Nuclear Structure Vol. II, Nuclear Deformations*, W. A. Benjamin, Reading, MA

Boltzmann, L., (2011). *Lectures on Gas Theory*, Dover Publications, Mineola, NY; originally published as *Vorlesungen über Gastheorie*, Leipzig, GR (1896)

Bose, S. N., (1924). *Z. für Phys.* **26**, 178

Chandler, D., (1987). *Introduction to Modern Statistical Mechanics*, Oxford University Press, New York, NY

Colorado, (2011). *The Bose-Einstein Condensate*, http://www.colorado.edu/physics/2000/bec

Darwin, C. G., and Fowler, R. H., (1922). *Phil. Mag.* **44**, 450, 823

Davidson, N., (2003). *Statistical Mechanics*, Dover Publications, Mineola, NY; originally published by McGraw-Hill, New York, NY (1962)

Debye, P. J. W., (1912). *Ann. der Phys.* **39**, 789

Debye, P. J. W., (1988). *The Collected Papers of P. J. W. Debye*, Ox Bow Press, Woodbridge, CT

Dennison, D. M., (1927). *Proc. Roy. Soc. (London)*, **A115**, 483

Dirac, P. A. M., (1926). *Proc. Roy. Soc. (London)*, **A112**, 664

Dubach, J., (2004). Private communication, quoted in [Walecka (2004)]

Edmonds, A. R., (1974). *Angular Momentum in Quantum Mechanics*, 3rd printing, Princeton University Press, Princeton, NJ

Einstein, A., (1907). *Ann. der Phys.* **22**, 180

Einstein, A., (1924). *Sitz. der Preuss. Akad. der Wiss., Phys.-Math. Klass*, 261

Fermi, E., (1926). *Rend. Accad. Naz. Lincei* **3**, 145

Fermi, E., (1927). *Rend. Accad. Naz. Lincei* **6**, 602

Fetter, A. L., and Walecka, J. D., (2003). *Quantum Theory of Many-Particle Systems*, Dover Publications, Mineola, NY; originally published by McGraw-Hill, New York, NY (1971)

Fetter, A. L., and Walecka, J. D., (2003a). *Theoretical Mechanics of Particles and Continua*, Dover Publications, Mineola, NY; originally published by McGraw-Hill, New York, NY (1980)

Fetter, A. L., and Walecka, J. D., (2006). *Nonlinear Mechanics: A Supplement to Theoretical Mechanics of Particles and Continua*, Dover Publications, Mineola, NY

Feynman, R. P., Metropolis, N., and Teller, E., (1949). *Phys. Rev.* **75**, 1561

Fowler, R. H., and Guggenheim, E. A., (1949). *Statistical Thermodynamics, rev. ed.*, Cambridge University Press, Cambridge, UK

Gibbs, J. W., (1960). *Elementary Principles in Statistical Mechanics*, Dover Publications, Mineola, NY; originally published by Yale University Press, New Haven, CT (1902)

Gibbs, J. W., (1993). *The Scientific Papers of J. Willard Gibbs, Vol. 1: Thermodynamics*, Ox Bow Press, Woodbridge, CT

Gross, D. J., and Wilczek, F., (1973). *Phys. Rev. Lett.* **30**, 1343

Gutiérrez, G., and Yáñez, J. M., (1997). *Am. J. Phys.* **65**, 739

Hartnack, C., Kruse, N., and Stöcker, H., (1993). *The Vlasov-Uehling-Uhlenbeck Model*, in *Computational Nuclear Physics Vol. 2*, eds. K. Lananke, J. A. Maruhn, and S. E. Koonin. Springer-Verlag, New York, NY p.128

Herzberg, G., (2008). *Molecular Spectra and Molecular Structure, Vol. I. Spectra of Diatomic Molecules*, Reitell Press, Paris, FR

Huang, K., (1987). *Statistical Mechanics, 2nd ed.*, John Wiley and Sons, New York, NY

Ising, E., (1925). *Z. Phys.* **31**, 253

Kadanoff, L. P., (2000). *Statistical Physics*, World Scientific Publishing Company, Singapore

Kiel, (2016). *http://www.theo-physik.uni-kiel.de/~bonitz/D/vorles_12ss/ising_data.pdf*

Kittel, C., and Kroemer, H., (1980). *Thermal Physics, 2nd ed.*, W. H. Freeman and Co., New York, NY

Kittel, C., (2004). *Introduction to Solid State Physics, 8th ed.*, John Wiley and Sons, New York, NY

Kohn, W., (1999). *Rev. Mod. Phys.* **71**, 1253

Kramers, H. A., and Wannier, G. H., (1941). *Phys. Rev.* **60**, 252

Kubo, R., (1988). *Statistical Physics: An Advanced Course with Problems and Solutions*, North-Holland, Amsterdam, NL

Landau, L. D., and Lifshitz, E. M., (1980). *Statistical Physics, 3rd ed.*, Pergamon Press, London, UK

Landau, L. D., and Lifshitz, E. M., (1980a). *Quantum Mechanics Non-Relativistic Theory, 3rd ed.*, Butterworth-Heinemann, Burlington, MA

Langmuir, I., (1916). *J. Am. Chem. Soc.* **38**, 2221

Lattice 2002, eds. Edwards, R., Negele, J., and Richards, D., (2003). *Nuclear Physics* **B119** (Proc. Suppl.)

Ma, S. K., (1985). *Statistical Mechanics*, World Scientific Publishing Company, Singapore

Mayer, J. E., and Mayer, M. G., (1977). *Statistical Mechanics, 2nd ed.*, John Wiley and Sons, New York, NY

McCoy, B. M., (2015). *Advanced Statistical Mechanics*, (International Series O Monographs on Physics) Oxford University Press, Oxford, UK

Metropolis, N., Rosenbluth, A., Rosenbluth, M., Teller, A., and Teller, E., (1953). *J. Chem. Phys.* **21**, 1087

Negele, J. W., and Ormond, H., (1988). *Quantum Many-Particle Systems*, Addison-Wesley, Reading, MA

Ohanian, H. C., (1995). *Modern Physics, 2nd ed.*, Prentice-Hall, Upper Saddle River, NJ

Onsager, L., (1944). *Phys. Rev.* **65**, 117

Pauli, W., (2000). *Statistical Mechanics: Vol. 4 of Pauli Lectures on Physics*, Dover Publications, Mineola, NY

Pauling, L., (1935). *J. Am. Chem. Soc.*, **57**, 2680

Plischke, M., and Bergersen, B., (2006). *Equilibrium Statistical Mechanics, 3rd ed.*, World Scientific Publishing Company, Singapore

Politzer, H. D., (1973). *Phys. Rev. Lett.* **30**, 1346

Reif, F., (1965). *Fundamentals of Statistical and Thermal Physics*, McGraw-Hill, New York, NY

Rushbrooke, G. S., (1949). *Introduction to Statistical Mechanics*, Oxford University Press, Oxford, UK

Schiff, L. I., (1968). *Quantum Mechanics, 3rd ed.*, McGraw-Hill, New York, NY

Stöcker, H., and Greiner, W., (1986). *Phys. Rep.* **137**, 277

Ter Haar, D., (1966). *Elements of Thermostatics, 2nd ed.*, Holt Reinhart and Winston, New York, NY

Thomas, L. H., (1927). *Proc. Cam. Phil. Soc.* **23**, 542

Tolman, R. C., (1979). *The Principles of Statistical Mechanics*, Dover Publications, Mineola, NY; originally published by Oxford University Press, Oxford, UK (1938)

Van Vleck, J. H., (1965). *The Theory of Electric and Magnetic Susceptibilities*, Oxford University Press, Oxford, UK

Walecka, J. D., (2000). *Fundamentals of Statistical Mechanics: Manuscript and Notes of Felix Bloch, prepared by J. D. Walecka*, World Scientific Publishing Company, Singapore; originally published by Stanford University Press, Stanford, CA (1989)

Walecka, J. D., (2004). *Theoretical Nuclear and Subnuclear Physics, 2nd ed.*, World Scientific Publishing Company, Singapore; originally published by Oxford University Press, New York, NY (1995)

Walecka, J. D., (2008). *Introduction to Modern Physics: Theoretical Foundations*, World Scientific Publishing Company, Singapore

Walecka, J. D., (2010). *Advanced Modern Physics: Theoretical Foundations*, World Scientific Publishing Company, Singapore

Walecka, J. D., (2011). *Introduction to Statistical Mechanics*, World Scientific Publishing Company, Singapore

Walecka, J. D., (2013). *Topics in Modern Physics: Theoretical Foundations*, World Scientific Publishing Company, Singapore

Wannier, G. H., (1987). *Statistical Physics*, Dover Publications, Mineola, NY; originally published by John Wiley and Sons, New York, NY (1966)

Wiki (2010). *The Wikipedia*, http://en.wikipedia.org/wiki/(topic)

Wilson, E. B., Decius, J. C., and Cross, P. C., (1980). *Molecular Vibrations: The Theory of Infrared and Raman Vibrational Spectra*, Dover Publications, Mineola, NY; originally published by McGraw-Hill, New York, NY (1955)

Wilson, K., (1971). *Phys. Rev.* **B4**, 3174

Wilson, K., (1974). *Phys. Rev.* **D10**, 2445

Yang, C. N., (1952). *Phys. Rev.* **85**, 808

Zemansky, M. W., (1968). *Heat and Thermodynamics: an Intermediate Textbook, 5th ed.*, McGraw-Hill, New York, NY

Index